Environmental Economics for Watershed Restoration

Environmental Economics for Watershed Restoration

Edited by
Hale W. Thurston
Matthew T. Heberling
Alyse Schrecongost

CRC Press
Taylor & Francis Group
Boca Raton London New York

CRC Press is an imprint of the
Taylor & Francis Group, an **informa** business

Cover photo courtesy of Annie Simcoe

CRC Press
Taylor & Francis Group
6000 Broken Sound Parkway NW, Suite 300
Boca Raton, FL 33487-2742

First issued in paperback 2017

© 2009 by Taylor & Francis Group, LLC
CRC Press is an imprint of Taylor & Francis Group, an Informa business

No claim to original U.S. Government works

ISBN 13: 978-1-138-11480-7 (pbk)
ISBN 13: 978-1-4200-9262-2 (hbk)

Library of Congress Cataloging-in-Publication Data

Evironmental economics for watershed restoration / edited by Hale W. Thurston, Matthew T. Heberling and Alyse Schrecongost.
    p. cm.
Includes bibliographical references and index.
ISBN-13: 978-1-4200-9262-2 (alk. paper)
ISBN-10: 1-4200-9262-6 (alk. paper)
    1. Watershed restoration--Economic aspects. 2. Watershed management--Economic aspects. I. Thurston, Hale W., 1965- II. Heberling, Matthew T., 1971- III. Schrecongost, Alyse. IV. Title.

TC409.E58 2009
333.73'153--dc22
                                         2008038891

Visit the Taylor & Francis Web site at
http://www.taylorandfrancis.com

and the CRC Press Web site at
http://www.crcpress.com

Contents

Preface

Traveling through West Virginia for pleasure and work, watching the beautiful hills, streams, hollows, and gullies pass by, one cannot help but admire the natural resources and environmental services available to the people who live there. For many West Virginians, their backdoor view is picturesque scenery of forested hills and streams, and their recreation destination of choice is likely the banks of a river or camping in the mountains. Unfortunately, in many areas past choices have caused damage to those resources, and current choices place the environment in danger of more damage. Because of the open access nature of rivers and streams, the role of watershed associations to protect and restore the beauty of a river or stream is critical. No one in particular owns the rivers or streams; they are public goods that have many uses that can be exploited. They can be used as conveyance for waste such as tailings from mines, as destinations for recreation, as providers of other ecosystem services, and much more. Watershed associations usually center their efforts on addressing pollution problems or helping to protect many of the various uses, but they face difficult choices. Often, the funds are just not available to address all the damages or threats. In some cases, watershed associations must advocate for new sources of funding to expand their activities or new legal protections to reduce threats. In other cases, they must justify their choices of protecting certain uses as opposed to others. None of these activities is easy.

When we consider the value of our streams for fishing or kayaking and other services such as cultural and existence values and the ecosystem services they provide, we can begin to understand the importance of restoration and protection. When evaluating the costs and benefits of doing a restoration project, all of these values have to be included on the *benefits* side of the equation. This book is intended to help quantify those benefits.

Although we wrote this book because we saw how important the roles of many stream restoration advocates and watershed associations are in bringing back some of the wild nature of Appalachia, the same decisions are being made throughout the country. We saw and heard about watershed groups struggling with some of the economic analyses they were being asked to do or that they knew they needed to do. This book is a compilation of economic approaches that address some of the many problems and choices faced by watershed groups and restoration advocates. In some cases, economic analysis does require the input from a trained, doctoral-level economist from a think tank or university, but in many cases the level of analysis needed to get a decent grip on the problem at hand, to do a back-of-the-envelope benefit-cost calculation, or to rank several proposed projects is something that can be tackled by someone in the watershed association or community starting with a good grasp of the watershed restoration problem at hand.

We are grateful to many who have helped bring this book to light. We would like to thank the people at Canaan Valley Institute for their insight and guidance, especially Ryan Gaujot for his geographic information systems knowledge, Paul Kinder for his kayaking prowess, Jenny Newland, Ron Preston, and Jim Rawson. We owe a special debt of gratitude to Dr. Dave Szlag, who suggested most of this in the first place. The staff and volunteers of the Friends of Deckers Creek deserve thanks for their continuous involvement and feedback. We also wish to thank Jennifer Ahringer and Marsha Hecht for their helpful assistance in manuscript preparation. We thank Dr. Heriberto Cabezas for his support of this work. Thanks to Ben Gilmer and Richard Herd for allowing this project to encroach on their project and activity space. Finally, we [Hale and Matt] thank our wives, Yngrid Thurston and Jacqueline Heberling, for their support.

About the Editors

Matthew T. Heberling is an economist in the U.S. Environmental Protection Agency's (USEPA's) National Risk Management Research Laboratory (NRMRL), Sustainable Environments Branch, in Cincinnati, Ohio. He holds a Ph.D. in agricultural economics from the Pennsylvania State University, where he specialized in environmental and natural resource economics. Dr. Heberling joined the USEPA in 2001 to work in a program of research integrating ecological risk assessment and economic analyses. He is now studying the effects of ancillary benefits on market mechanisms and conducting research on sustainability metrics. His research experience also includes using economic valuation methods to examine individuals' preferences for recreational fishing and to prioritize stream restoration. When he visits his hometown in Pennsylvania, he enjoys fishing for smallmouth in the Susquehanna River.

Alyse Schrecongost is a natural resource economist currently working as an agricultural development analyst for the Bill and Melinda Gates Foundation's Science and Technology Initiative. Ms. Schrecongost has been working on natural resource management issues through policy work, community development, and academic research for the past 10 years, with a specific focus on institutional economics and water quality management. Her work experience ranges from researching irrigation systems in West Africa to developing a water quality trading program in the Potomac River basin of West Virginia. Related projects have addressed water gauging networks, urban storm water institutions, and landfill methane capture. Ms. Schrecongost's travels have allowed her to research watershed management recreationally in Cuba, Morocco, Sri Lanka, and other interesting places. A West Virginia native now living in the Pacific Northwest, she and her husband spend a lot of time in state and national parks, appreciating historical efforts to restore and protect natural areas.

Hale W. Thurston is an economist in the USEPA's NRMRL, Sustainable Environments Branch, in Cincinnati, Ohio. He received his Ph.D. in economics from the University of New Mexico, a master's degree in international affairs from Ohio University, and a bachelor's degree in English literature from Bates College. His research currently focuses on nonmarket valuation of natural resources in the policy arena and the use of market incentives to promote low-impact development.

Contributors

Stephen Beaulieu
Research Triangle Institute
Research Triangle Park
North Carolina

Elena Y. Besedin
Abt Associates Inc.
Cambridge, Massachusetts

Randall J.F. Bruins
National Exposure Research Laboratory
U.S. Environmental Protection Agency
Cincinnati, Ohio

Alan R. Collins
Agricultural and Resource
 Economics Program
West Virginia University
Morgantown, West Virginia

Jonathan I. Eisen-Hecht
ICF International
Fairfax, Virginia

Jerald J. Fletcher
Agricultural and Resource
 Economics Program
West Virginia University
Morgantown, West Virginia

Evan Hansen
Downstream Strategies LLC
Morgantown, West Virginia

Matthew T. Heberling
National Risk Management Research
 Laboratory
U.S. Environmental Protection Agency
Cincinnati, Ohio

Robert J. Johnston
George Perkins Marsh Institute
Department of Economics and
Clark University
Worcester, Massachusetts

Randall A. Kramer
Nicholas School of the Environment
 and Earth Science
Duke University
Durham, North Carolina

Randall S. Rosenberger
Department of Forest Resources
Oregon State University
Corvallis, Oregon

Alyse Schrecongost
West Virginia Water Research Institute
West Virginia University
Morgantown, West Virginia

Anne Sergeant
National Center for Environmental
 Research
U.S. Environmental Protection Agency
Washington, DC

Hale W. Thurston
National Risk Management Research
  Laboratory
U.S. Environmental Protection Agency
Cincinnati, Ohio

George Van Houtven
Research Triangle Institute
Research Triangle Park
North Carolina

Gene E. Vaughan
Duke Energy Corporation—
  Environmental Center
Huntersville, North Carolina

James M. Williamson
National Risk Management Research
  Laboratory
U.S. Environmental Protection Agency
Cincinnati, Ohio

Abbreviations

AMD: acid mine drainage
AR: autoregressive
BCA: benefit-cost analysis
BT: benefits transfer
CEA: cost-effectiveness analysis
CPI: consumer price index
CVI: Canaan Valley Institute
CVM: contingent valuation method
CWA: Clean Water Act
EIA: economic impact analysis
FODC: Friends of Deckers Creek
GIS: geographic information systems
GLM: generalized linear model
HUC: hydrologic unit code
I/O: input/output analysis
IMPLAN: Impact Analysis for Planning
MA: Millennium Ecosystem Assessment
NAIC: North American Industry Classification
NOAA: National Oceanic and Atmospheric Administration
NRCS: Natural Resource Conservation Service
O&M: operation and maintenance
RFF: Resources for the Future
RPC: regional purchase coefficients
SAM: social accounting matrix
SCM: stated choice method
TCM: travel cost method
TMDL: total maximum daily load
USEPA: U.S. Environmental Protection Agency
WARMF: watershed analysis risk management framework
WCAP: Watershed Cooperative Agreement Program
WQS: water quality standards
WTP: willingness to pay
WV-DEP: West Virginia Department of Environmental Protection

1 Introduction to Economic Jargon and Decision Tools

Hale W. Thurston, Matthew T. Heberling, and Alyse Schrecongost

CONTENTS

INTRODUCTION

To manage something it must first be measured. When dealing with options for restoring ecosystems or water bodies, often we want to measure the value of undertaking a certain project. Estimating the value of a restoration project can help us to prioritize projects when budgets are limited. A watershed association may want to determine the costs and benefits of stream restoration or other stream-improving activities. This might be because they want to compare two or more potential projects to determine where to spend their money. It may be because they have been asked by people in the community or policy makers to justify their request for funds to carry out a restoration project. The community may want to know what they are giving up (e.g., new playground) for improved water quality.

Usually, determining the costs of projects is easier than determining benefits because these costs are based on things sold in the market.[1] The costs of, say, holding an annual "Clean-up the Such-and-Such Watershed" canoe outing are relatively easy to estimate. Canoe rental plus cost of plastic garbage bags, plus maybe a per

1

hour estimate of volunteers' time, plus the cost of sodas and hamburgers consumed at the picnic afterward, plus the dump fee add up to the total cost of the event. One might want to exclude the cost of volunteer time because everyone had fun, and one might want to add some other costs, perhaps a picnic area daily fee. It is not an exact science, but as long as things are itemized, the total is "transparent" or justifiable. Even the costs of some larger-scale restoration projects are relatively easy to figure out as long as one is meticulous about adding all factors.

The *benefits* of restoration projects, however, while obvious to the green-oriented member of an environmental group, are not as easily quantified. By *quantified*, we mean monetized or putting a dollar value on the benefits. In addition, we refer to monetizing the benefits as *valuation*, and doing these valuation studies is not so straightforward. This is why we developed this book. Valuation is usually done in an academic setting or by highly paid consultants. It is our position that a thoughtful layperson can, with the right tools, perform at least a preliminary estimate of some potential benefits of stream restoration projects. Chances are if you are reading this book it is probably because you have encountered policy makers or funding organizations who want to know how much bang they are going to get for their restoration buck. They may not even want to hear details about how ecological conditions in the stream are going to improve; they may be more interested in a dollar value of those changes.

We should also take this opportunity to mention that economics is only a part of the equation. There are many occasions when an overriding ecological condition, such as the existence of critical habitat for an endangered species, will automatically justify a certain project or make it the first priority for restoration. Other considerations besides economic valuation can play an equal role in decision making. This may be as simple as the geographic location of a reach of stream that makes it the most attractive of many to restore first, some local cultural value of a body of water, or politics. If necessary, some of these things (ease of access, historical value, or ethics) are occasionally valued by economists, but the methods themselves are costly and intricate.

Why, when, and how to use environmental economics in watershed project analysis can be confusing or even disturbing. The intent of this book is to provide guidance to watershed groups interested in understanding or even incorporating economic valuation for prioritizing many projects or to justify spending a certain amount of money on a project. This book does not replace the services of a trained economist in most cases, but it should provide a basic background on the types of ecological goods and services (i.e., the ecological functions and processes that directly or indirectly affect an individual's well-being or satisfaction, like water purification or flood control) that are often valued and the types of questions that should be asked. It should make stakeholder groups more comfortable talking about things like contingent valuation, marginal costs, nonmarket goods, and other economic jargon.

Economic Jargon

By way of introduction, we would like to begin where many stakeholders probably start. On being asked to come up with monetary values for the benefits of some restoration project or projects, they might start by looking at an environmental economics textbook or academic journals for guidance. Here, taken almost at random and admittedly out of context, from a recent book on environmental valuation, is an example of what you are up against in the economics literature:

> A stated preference model was estimated using maximum likelihood. All attributes were included in the model as were two alternative specific constant (ASCs) (one for each hunting alternative). The alternative of not hunting (non-participation) is the third alternative and attribute levels are assumed to be zero for this choice. Adamowicz et al. (1997, p. 72)

This passage about a moose-hunting study is written by highly respected economists, and the study is well done, but it is unreasonable to expect an expert in community organizations or watershed groups to fully understand it. Our book communicates the information and important points in this passage to noneconomists in a language that is less foreign. We would like readers to know where the authors are going with such a study and to determine if such a study is important to your watershed work. In this chapter, we introduce many of the concepts that sound like gibberish to the untrained ear. While you might disagree with some of these definitions, it is nevertheless important to know that they are the generally accepted definitions in the field of economics. Most of the definitions can be found in the Glossary. In addition, we have created text boxes throughout the book that either expand on a definition or concept highlighted by or point out difficult concepts using ⚠ .

Goods

When we mention goods, you might think of something that you value and can be purchased in a market. It turns out that a good can be many things. For example, a tube of toothpaste is a good, but there are also environmental goods like clean water. Economists talk about goods as *exclusive* or *nonexclusive* and *rival* or *nonrival* in how they are consumed or enjoyed. *Exclusive* asks the question: are some individuals prevented from benefiting from a good or service once produced? *Rival* is determined by asking: does one person's enjoyment of a thing reduce the enjoyment others can get from it? Private consumer goods like pizza and beer are rival and exclusive. These consumer goods, because of their characteristics, can be traded in markets. Markets can be studied and in turn allow economists to determine the financial value of the goods.

CATEGORIES OF GOODS

Rival—if one person uses it, another person cannot.
Nonrival—if one person uses it, another person can still use it.

Exclusive—people can be kept from enjoying the good if they do not pay for it.
Nonexclusive—people cannot be kept from enjoying the good even if they did not pay for it.

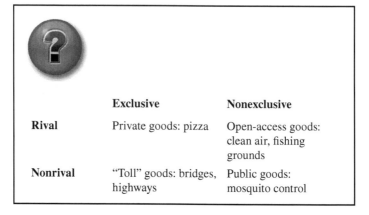

	Exclusive	Nonexclusive
Rival	Private goods: pizza	Open-access goods: clean air, fishing grounds
Nonrival	"Toll" goods: bridges, highways	Public goods: mosquito control

Most goods of concern in watershed restoration are rival (usually due to crowding, but also to pollution) and nonexclusive (because healthy fisheries cannot easily be fenced in). These are the typical *open-access goods*. A public fishing area, for example, is rival to an extent because there are fewer fish and less elbow room to go around if there are more people in the area and nonexclusive because no one is denied access to the benefits of the good. Other techniques are needed to value nonmarket goods and services or those that are not traded in markets.

ECONOMIC AND FINANCIAL VALUES

Watershed groups and other environmentally concerned groups can be frustrated by environmental economics because it is the business of associating dollar values to (*monetizing*) natural resources that seem to be invaluable. Environmental economists, however, recognize that while some things are beyond a financial valuation, many environmental goods have very low financial value but very high economic value. It is their job to account for the many values of a good that are not captured by existing financial markets like real estate markets and the pollution treatment technology costs. *Economic value* is the net value of a good to the public minus the financial or market cost of providing or protecting that good.

This discussion is simplified greatly to avoid confusion over economists' definitions of value. Understand that when you see consumer's surplus that this is an approximation of the true change in well-being or value. For a detailed theoretical discussion of the true measure of welfare change, see Freeman (2003).

A good has an economic value if it matters to people—if it affects an individual's perceived or actual well-being. Our level of well-being is related to the level of satisfaction we get from an action or activity. Economists talk about value in terms of trade-offs or gains and losses of actions, activities, or behavior. People's *willingness to pay* or *willingness to accept* are measured to estimate economic value. For example, if a stream is restored, a person would be willing to pay *x* amount of money and consider himself or herself as well off as before the improvement was made. If a stream is degraded, an individual would be willing to accept *x* amount of money as compensation for the loss to keep the personal level of satisfaction the same. This is a common sticking point between economists and ecologists. Ecologists point out that things like clean air and bird habitat have value on their own, which is different from the perspective that it has to matter to people. The economic value of things has to be seen as only one measure of a good's total intrinsic value.

CATEGORIES OF ECONOMIC VALUES

Use Value

Direct use—fish in stream to catch and eat.

Indirect use—shade trees and clean water required for healthy fish habitat. Wetlands that provide flood control or water filtration.

Nonuse Values

Existence/option—stream stretch that provides spiritual value to community group or population or has historical significance.

Bequest—value in preserving natural asset for the enjoyment of future generations.

Economists go on to distinguish types of value. Broadly, these are *use value* and *nonuse value*. Goods are further broken down to have direct or indirect value. Goods, including open-access goods, have use value because they are "consumed" or enjoyed directly. One might directly value fishing from a clean mountain stream; indirectly, we may value trees along a stream bank because they provide shade for trout habitat.

Nonuse values are different and harder to put your finger on. Nonuse values for environmental goods can, however, be very large. These include *bequest value,* which is a value those of us in this generation have for preserving an environmental good for the benefit of future generations, and *existence value*, which is the value associated with just knowing that a good exists and getting some well-being out of that knowledge. The example usually given is the fact that many people value whales enough to periodically send money to Greenpeace even though many of those people never plan to actually see a whale.

Why place a value on things like natural services at all? One reason is because it has been pointed out that "what is not managed is often neglected," and one aspect of management is valuation of what you have. Another reason for valuation is less esoteric—it often happens that we are faced with project prioritization under some kind of budget constraint. There just is not enough money to restore every stream affected by acid mine drainage in Appalachia.

Economists can approach prioritization of restoration projects using different analyses or some combination of analyses, depending on the objective of the study (e.g., project design, project justification, project prioritization, etc.). They can examine the *efficiency* of projects, determining whether the total benefits are larger than the total costs to society of not doing the project. If the *distribution* of costs and benefits from the project is important, then *economic impact analysis* (EIA) helps identify the project's winners and losers within affected economic sectors or *equity assessment* helps identify the effects to subpopulations of interest.

DECISION TOOLS

Before we describe how to estimate benefits and costs, we present frameworks that support watershed decisions. In this way, you will have a better understanding of why economists are interested in the benefits and costs. We detail some of the formal decision tools that are used to support decisions. You might not know about some of the more formal aspects of them and some you might not have heard about before now.

Benefit-cost analysis (BCA) is a commonly used tool for decision makers. While BCA is not exact, it can be an accurate, transparent exposition of the majority of the costs and benefits a project is likely to incur/reap over its lifetime. It causes those who are interested in the project to delineate several aspects of the project and can help discover some costs or benefits that might have gone unrecognized had the BCA not been considered.

In a certain sense, BCA is just a term for one of many decision tools that we use all the time in our daily lives: we consider the option of doing a thing and think about what it will get us (benefits) and what we have to give up for it (costs). In the policy world, it is more structured but the same in spirit. A BCA measures the net gain or loss to society at large due to a certain policy or project. It can also be used to compare two or more different options available to us, as we discussed in this chapter. BCA is used to compare or measure projects of all kinds: freeway widening, sewage plant building, manufacture plant sighting, and of interest in the current situation, it is often used to compare proposed watershed restorations. It provides the policy maker with a transparent list of the various and many pros and cons (benefits and costs) of any large project before deciding yea or nay. The biggest knock on BCA is that it distills a whole project down to one number, and that it may not be the decision-making tool one wants to use. There are other ways to decide on potential undertakings. Traditionally, a complete economic analysis augments a BCA with an EIA, and an equity assessment (U.S. Environmental Protection Agency [EPA] 2000). BCA provides information about economic efficiency; the other two techniques examine resource distribution. King (2005) correctly noted some of the essential shortcomings and nuances of BCA:

> BCA is only one of many possible ways to make public decisions about the natural environment. Because it focuses only on economic benefits and costs, benefit-cost analysis determines the economically efficient option. This may or may not be the same as the most socially acceptable option, or the most environmentally beneficial option. Remember, economic values are based on peoples' preferences, which may not coincide with what is best, ecologically, for a particular ecosystem. However, public decisions must consider public preferences, and benefit-cost analysis based on ecosystem valuation is one way to do so. Often, when actual decisions are made, a benefit-cost analysis will be supplemented with other information, such as equity implications or overriding environmental considerations.

There are essentially five steps to any BCA; depending on how one breaks it out, this could be fewer or more, but the following elements are critical:

1. *Define the proposed project.* Explicit in this very important step is the delineation of the policy area. To provide legitimate estimates, a BCA requires painstaking delineation of the study area, including geographic scale, demographic extent, and time frame. What group of people will be affected? Furthermore, BCA relies on neoclassical economic underpinnings as such a distinct change in an environmental amenity that must be defined.
2. *Identify the impacts of the project, both positive and negative.* Specifically, what will the restoration do? Create better habitat? Impinge on private property? Increase tourism? Decrease tourism?
3. *Quantify the impacts.* Determine the technical effectiveness from engineering and ecological studies. Just putting riprap in a stream bend does not totally stop erosion. We need to know how much it helps both where it is and what it will do for the stream as a whole.

   Assume or estimate performance endpoints from the ecology literature. If the erosion is reduced by 50%, what does that mean to fish or other critters downstream? This is important if we are going to find out how people value the restoration project; for people to value things, they need to know how the project affected things in the stream that they are aware of: fishing, sightseeing, rafting, biodiversity, and the like.
4. *Estimate costs and benefits.* Usually, determining the costs of projects is easier than determining benefits because the costs are based on things sold in the market, but we need to be sure we are not omitting relevant costs. The focus of the remaining chapters is on the different approaches for quantifying the benefits from watershed restoration. The next chapter goes into more detail on the different approaches for monetizing the change in environmental quality. These techniques include travel cost method, hedonic pricing method, contingent valuation method, stated choice method, and benefit transfer.
5. *Discounting.* Often, the costs and benefits occur in different periods in the life of a project. Costs are usually borne immediately, while especially

in the case of a natural system that needs time to grow, benefits accrue later in the life of the project. Because people have preferences for the time value of money, these differences need to be accounted for (USEPA [2000] provides a detailed explanation of discounting).

A Hypothetical BCA Should Help Bring All This Together

The watershed group Friends of the Spoon River wants to undertake an acid mine drainage remediation project. They have identified two potential projects: Project 1 involves building a limestone drum facility that will raise the pH in the main stem of the Spoon from 2.5 to 7 for 12 miles. Project 2 calls for the installation of seven 100-meter limestone channels along five first-order tributaries to the Spoon, raising the pH in the tributaries from 2 to 7 for 10 miles. The drum facility costs $300,000 with a lifetime of 7 years. Operation and maintenance (O&M) figures to be $2,500 per year. The pH increase allows for fishing after 2 years of operation, and 3,000 more fishing days per year (e.g., 1,000 anglers fishing 3 days per year) would occur due to the project. The group uses the travel cost method to estimate that a fishing day is worth $25 to the local economy. Option 2 uses the lower-technology limestone channels, which cost $300 per 50 meters. The channels also last for 7 years, and O&M is $50 per year per 50 meters. Although the tributaries will support life after the channels are installed, they probably will not be fished recreationally except by a handful of locals who have been fishing other small streams. The increase in macro-invertebrates, however, within a year would improve stream and riparian area health and appearance. Using the hedonic pricing method, the group reckons the value of the 18 houses in the immediate area would increase by $500 on average. The various costs and benefits are presented in Table 1.1. For expository purposes only, we include nondiscounted figures on the left, and the choice of 5% as a discount rate is ad hoc (USEPA [2000] suggested a 2% to 3% discount rate and a 7% discount rate for their analyses).

One way to compare BCA figures is to create a ratio of the benefits to the costs. A number greater than 1 means, of course, benefits are more than costs. When the ratio is greater than 1, we say that the project or action increases efficiency. A few interesting things come out of this exercise. The benefit–cost ratio of Project 1 (312,293/298,044) is 0.95, while the benefit–cost ratio of Project 2 (4,691/8,490) is 1.81. Notice that Project 2 is much better according to the ratios, but what happens if costs and benefits are not discounted? The reason behind this is that discounting reduces future benefits and costs. The large benefits from Project 1 that occur in the later years do not matter as much, but when they are not discounted, they are on equal footing with present benefits and costs.

Economic Impact Analysis

Another approach to valuing ecological changes is to look specifically at the change in the local economy from an increased demand in outdoor recreation or tourism, such as from fishing or bird watching. The increased demand may be caused by improved environmental quality or advertising to those outside the local area. EIA is

TABLE 1.1
Benefit–Cost Analysis for Spoon River Restoration

	Project 1		Project 2		Project 1: 5% Discount Rate		Project 2: 5% Discount Rate	
	Costs	Benefits	Costs	Benefits	Costs	Benefits	Costs	Benefits
Year								
0	300,000	0	4,200	0	300,000	0	4,200	0
1	2,500	0	100	9,000	2,358	0	94	8,490
2	2,500	75,000	100	0	2,224	66,749	88	0
3	2,500	75,000	100	0	2,099	62,971	83	0
4	2,500	75,000	100	0	1,980	59,407	79	0
5	2,500	75,000	100	0	1,868	56,044	74	0
6	2,500	75,000	100	0	1,762	52,872	70	0
Total	315,000	375,000	4,800	9,000				
Net present value					312,293	298,044	4,691	8,490

the technique used to measure these changes, and input/output (I/O) modeling is an important component of this analysis. Following is a brief review.

Input/Output Analysis

I/O analysis examines the interconnection of industries in the region to show how changes in final sales affect the regional income of each industry. It can simply be a descriptive tool or a predictive tool. Using I/O analysis, we can predict the total impact of a small change in an economic activity (say, an increase in outdoor recreation).

The basic objective of this analysis is to examine the effects of a new industry, expansion of a firm, a cutback in government spending, or a public investment venture. The change in activity will cause a ripple (or multiplier) effect until the economy is in equilibrium again. The sum of direct, indirect, and induced impacts equals the total impact on an economy. For outdoor recreation, the direct expenditures are those that remain in the area and may include motels, transportation, food, or outdoor equipment. Indirect expenditures include the firms that benefit from the visitors' expenditures, like guides and outfitters. These firms must buy goods, services, and inputs (hopefully from local suppliers). Indirect expenditures create income throughout the local economy and maintain the employment (Crabtree et al. 1994). The increase of local households' spending generates induced expenditures. This occurs from the recreationists' spending and increased wages, salaries, local profits, and rents (Crabtree et al. 1994).

All this "multiple spending" is captured in what is called a *multiplier*. I/O analysis provides measures of economic impacts through output, employment, and value-added multipliers. It is possible to estimate multipliers for other factors if the data exist and are linked to the output. For example, water use and pollution multipliers have been estimated (Horton 2001). Understanding a community's economic base and regional income and the interactions of an economy form the basis of the multipliers. The community can use this information to determine how a change in basic employment or a change in spending would benefit or hurt the community.

Economic planners can use the multipliers to detect the ripples caused by changes in employment or income. But, there are some concerns about using multiplier analysis. Although the multiplier concept seems like a good idea for planners and analysts, they must take caution when using multipliers. Many assumptions have been used to develop the whole chain, and any breakdown in those assumptions will cause problems in the interpretation (Richardson 1979).

There are also some difficulties when using the I/O analysis with tourism or recreation (Fletcher 1989). For example, one assumption is that no pollution or no environmental degradation occurs. In most cases, this is not true. Anyone who has visited a popular campground after a weekend knows there is usually some cleaning up to do. Another assumption is "constant returns to scale," which means that when demand causes output to increase, then inputs will increase proportionately. In the case of recreation, how could a recreation area like a state park increase to match demand?

I/O is one kind of economic impact study used to estimate changes in an economy. IMPLAN is computer software that does I/O analysis. IMPLAN stands for IMpact analysis for PLANning, and it was originally developed for the U.S. Forest Service to assist in its management planning (Minnesota IMPLAN Group 2000).

Bergstrom et al. (1990) estimated the regional impacts of outdoor recreation using IMPLAN. They obtained their visitor spending data from the Public Area Recreation Visitors Study. Visitors to state parks in North Carolina, South Carolina, Georgia, and Tennessee were surveyed to calculate multipliers. Overall, Tennessee had the largest multipliers, while South Carolina had the smallest (output, income, and employment). Output multipliers ranged from 1.80 to 2.46, while income multipliers ranged from 2.01 to 2.83 (i.e., this suggests that for each dollar of income generated by the state park, 1.83 dollars of income are created). Employment multipliers ranged from 1.36 to 1.81. Most of the multipliers were larger than expected (1.20 and 1.50). The response rates for the different states varied from 22% to 45%, which may have caused some biases, however.

If decision makers are interested in the effects on the local economy, this type of analysis will provide estimates of the effects. However, many other impacts are lost in this analysis, suggesting other types are needed. Management alternatives will usually change the baseline characteristics in some ways either positively or negatively but might leave others unchanged. BCA can reveal efficient alternatives but cannot really describe the winners and losers. EIA can tell us who the winners and losers are given the management alternative. Therefore, only looking at the results of IMPLAN will only provide a narrow perspective.

I/O can provide information at different scales of the economy. However, as the scale gets smaller, the impact will most likely become extremely small and difficult to interpret. The data may also not exist for small regions such as a small town with few industries. These are issues that need to be kept in mind when designing a research plan.

EQUITY ASSESSMENT

So far, we have discussed how to measure changes in efficiency with BCA and how to measure the economic impact using I/O. Although efficiency and economic impact matter for whether it makes sense to move forward with a project, it does not say anything about who receives the benefits and who faces the costs. Some projects may lead to a select few receiving the benefits while the entire community is left to pay the costs. Although it may be efficient or lead to improvement in economic impact, it may not be seen as fair.

Equity assessment tries to reveal who are the winners and losers of any particular project and typically focuses on subpopulations. In fact, USEPA (2000) thinks the net gains or losses or economic impact for vulnerable or disadvantaged subpopulations should be analyzed. One issue, however, is finding the data to examine the effects on subpopulations (USEPA 2000).

COST-EFFECTIVENESS ANALYSIS (CEA)

The last "framework" for supporting decisions is mentioned here because of the difficulties that can be found with BCA. A cost-effectiveness analysis (CEA) only examines the costs in dollar terms; the benefits are in nonmonetary units. It is typically used when quantifying the benefits is too difficult. The benefits may also be expressed as some goal or target, like amount of pollution to be reduced or a specific index of human health. When this is the case, CEA reveals the least-cost approach to meeting the goal. When different alternatives lead to different changes in the benefits, a ratio is calculated that suggests the number of dollars per unit of pollution reduced or per death avoided. In this way, the decision makers have information that ranks the alternatives.

SUMMARY

We have defined broad types of analyses in this chapter. Each can provide information to policy makers about how projects will change the current situation. EIA tells the decision maker how different sectors of the economy will change—whether they will grow or shrink. BCA, which organizes the changes to reveal something about efficiency, provides the decision maker information on whether the project is worth the effort. Equity assessment reveals how subpopulations gain or lose from a particular alternative. Each delivers different data and depends on what type of information the decision maker wants. We have presented the basic framework for supporting particular decisions. The next step is to describe the methods used to calculate the benefits or costs. The next chapter goes in depth into those valuation methods, including examples from actual studies.

Many details, problems, and solutions are brought up in the remaining chapters. Becoming familiar with these tools and concepts is the first step toward being able to discuss the economic importance of your project or watershed. We look at case studies that have many generalizable problems and solutions. Three of the five case studies focus on Appalachia and acid mine drainage problems. We focus on acid mine drainage because it is a major source of stream and river damage throughout the United States, affecting 32 states and thousands of miles of water ways (U.S. Geological Survey [USGS] 1998; USEPA 2002). After presenting the case studies, we present a technique that can be used as a communication tool for the community. Conceptual models are illustrations that connect ecosystems and their functions to goods and services that matter to individuals. By adding or subtracting pollution or restoration techniques, the conceptual models can illustrate how the ecosystems, their functions, and goods and services are affected. The last chapter describes how one watershed group utilized some of these economic methods to justify the additional funds needed to restore Deckers Creek in West Virginia. Because no two environmental problems are identical, these case studies serve as general guidance, not specific counsel. As mentioned, this book does not take the place of a professional economist, but we try to point out those tasks in the analysis that might require consulting services. We do not want this to read like a textbook. Although it is necessary to provide some initial definitions, we want this to be a reference book that is not unpleasant to read. We have tried to make it specific enough to be interesting and general enough to aid in a range of situations. We hope it is a useful reference to aid in environmental management decisions. A glossary is provided so that definitions can be looked up quickly.

NOTE

1. Of course, this is based on improving a water body or ecosystem. Watershed associations can also weigh the benefits and costs of some development project where development would lead to increased economic activity (i.e., benefits), but the costs would include the lost values of the stream (e.g., recreational fishing or aesthetics).

REFERENCES

Adamowicz, W., J. Swait, P. Boxall, and J. Louviere. 1997. Perceptions versus objective measures of environmental quality in combined revealed and stated preference models of environmental valuation. *Journal of Environmental Economics and Management*, 32(1): 65–84.

Bergstrom, J., H. Cordell, A. Watson, and G. Ashley. 1990. Economic impacts of state parks on state economies in the south. *Southern Journal of Agricultural Economics* 22(2): 69–77.

Crabtree, J., P. Leat, J. Santarossa, and K. Thomson. 1994. The economic impact of wildlife sites in Scotland. *Journal of Rural Studies* 10(1): 61–72.

Fletcher, J. 1989. Input-output analysis and tourism impact studies. *Annals of Tourism Research* 16(3): 514–524.

Freeman, A.M., III. 2003. *The Measurement of Environmental and Resource Values: Theory and Methods.* 2nd ed. Washington, DC: Resources for the Future.

Horton, G. 2001. Economic impact analysis: Assessing the effects of economic impacts: The derivation and application of economic, fiscal, resource and environmental impact multipliers. Division of Forecasting and Economic Impact Analysis. Nevada Division of Water Resources. Carson City, NV. http://www.state.nv.us/cnr/ndwp/forecast/econ_pg.4.htm.

King, D. 2005. Applying ecosystem value estimates—benefit cost analysis. Accessed October 2008 from www.ecosystemvaluation.org.

Minnesota IMPLAN Group, Inc. *IMPLAN Professional*. Version 2.0: User's Guide, Analysis Guide, and Data Guide. June 2000.

Richardson, H. 1979. *Regional Economics*. Urbana, IL: University of Illinois Press.

U.S. Environmental Protection Agency (USEPA). 2000. *Guidelines for Preparing Economic Analyses*. EPA 240-R-00-003. Washington, DC: Office of the Administrator.

U.S. Environmental Protection Agency (USEPA). 2002. Mid-Atlantic acidification. http://www.epa.gov/region03/acidification/ (accessed May 6, 2003).

U.S. Geological Survey (USGS). 1998. New hope for acid streams. Fact sheet, April. U.S. Department of Interior, U.S. Geological Survey, Biological Resources Division. http://www.lsc.usgs.gov/FactSheets/amdpub.pdf (accessed February 1, 2007).

2 A Closer Look at Valuation Methods and Their Uses

Hale W. Thurston, Matthew T. Heberling, and Alyse Schrecongost

CONTENTS

INTRODUCTION

In the hypothetical benefit-cost analysis (BCA) in Chapter 1, Friends of Spoon River determined that two options were available for their restoration. The benefits were estimated for each option, but you may ask specifically how could those numbers be calculated? How can the restoration have monetized benefits? Economists use two broad categories of preference observation—revealed and stated preference techniques—to estimate the economic values of restoration. These techniques and the variations of how economists could implement such studies to estimate watershed restoration or protection projects are described in this chapter.

REVEALED PREFERENCES

Broadly, economists measure preferences based on how people trade off money for other things. We can collect a lot of data on how people trade money for private consumer goods (remember rival and excludable) like toothpaste and pizza. These preferences are *revealed*; that is, they are expressed through certain buying behavior to someone like an economist or a marketing expert who is interested in such things. When environmental goods or services are not traded in markets, we can estimate values based on related market goods or services using revealed preference methods. We observe people's behaviors in certain markets and can infer estimated values for the environmental improvements that are not for sale in the marketplace.

TRAVEL COST METHOD

Revealed preference methods include noticing how much people spend to travel to a recreation spot; this gives us some idea of what the spot is worth to the person, even though the person might not expressly apply a money value to the spot. Because people choose to recreate at a particular location, we assume it has value, and their behavior (e.g., travel time and time spent at site) suggests its worth. Better still for an economic experiment, we might want to look at two outdoor recreation spots that are very similar but that might have one crucial difference, a better quality stream nearby, for example, and notice how much more people are "willing to pay" to go to the better stream site. This would give us an idea of the value of improving the stream at the worse site. A large amount of data are collected about people who visit a site and how much they spent to travel to some sites that differ in a special aspect; then, average amounts paid are estimated.

AVERTING BEHAVIOR, DEFENSIVE EXPENDITURES, AND REPLACEMENT COST

The terms averting behavior, defensive expenditures, and replacement cost all describe valuation techniques used by economists and ecologists that center on substitutability of new techniques or technology for the ecosystem services. For example, if the natural meander and riparian zone of a stream are lost due to construction, causing a need for levees, sump pumps, and sandbags to prevent flooding, then the money spent on the levees and other costs represent some valuation of the lost flood control ability of the natural stream system. Conversely, if the stream were restored, the amount of money households in the community would save by no longer needing to buy flood prevention items is a measure of the economic value of the restored system. This method, however, does not account for any nonuse values, only specific use values. Sometimes, economists argue that costs are not an appropriate measure of the benefits, so this method should be used with caution.

HEDONIC PRICING METHOD

Another method for valuation that takes advantage of people revealing their preference through market activity is the hedonic pricing method. This method relies on the existence of data on transactions, usually house sales, which are separated into various components of the price of the good. If two houses have the same number of bathrooms, the same number of bedrooms, the same view, the same everything except that one is overwhelmed with a pig farm odor, we can estimate how that pig smell is valued (surely negatively) even though it is never traded in a market. Since there are always some differences, we can use large samples of houses sold to call certain houses "statistically" identical and make that clear in our results. Similar to the travel cost method, we can develop an economic experiment to compare similar neighborhoods with different water quality or compare housing prices for the same neighborhood if water quality changes over time, say before and after a restoration project.

STATED PREFERENCES

Another general way to note preferences is simply to ask people what they would be willing to pay for a certain nonmarket good or service. We usually use stated preference methods when the good or service is not consumed often (that is, it has mostly "nonuse" value), and as such there is no record of transactions (like traveling to it or buying a house near it). The data on people's "willingness to pay" (WTP) are collected through a survey method following some standard rules of data collection, such as many outlined most famously in a book by Dillman (2000) called *Mail and Telephone Surveys: The Total Design Method*. This book explains the importance of polling a random sample, some of the methods for surveying (in person, phone, mail), and other important elements of conducting a defensible survey. More recently, the Internet has been used for eliciting preferences to reduce costs, but it still requires standard rules as well (see Thurston 2006; Dillman 2000). The stated preference questions are included in the survey in a special way, as we see in the outlines of the contingent valuation method (CVM) and the stated choice method (SCM) discussed in this chapter in more detail and in the case study section.

THE CONTINGENT VALUATION METHOD

The CVM is a way economists estimate use and nonuse values. The CVM involves lengthy surveys of people asking them to place a value on a change in a certain non-market good or service. There is usually some information in the survey about the good (e.g., a recreational fishing site could be described in detail along with potential changes to the quality of fishing, including expected number of fish caught, expected number of other anglers, etc.), and surveys are subject to approval and modification through the use of focus groups. There are also many biases that economists are aware of and try to minimize. If people understand that their responses could be used to help or hinder their situation, they may bias their true response up or down, depending on their situation. Remember, respondents are being asked to value a change in the nonmarket good. This is important because the theories behind modern economics do not allow for total valuation of a thing, only a "marginal" or small additional change in the thing. CVM is probably the most popular state preference method, but another approach is drawing interest.

THE STATED CHOICE METHOD

The SCM is another survey-based valuation technique that can be used to get at use and nonuse values. The SCM needs to adhere to all the same specifications for good surveys as CVM does, but the SCM tries to mimic a trade-off by giving the respondent to the survey a choice between several options that are described by their characteristics (e.g., in the quoted article in Chapter 1 about moose hunting in Alberta, Canada, the options dealt with hunting sites, and the characteristics to describe the sites included moose populations, hunter congestion, hunter access, forestry activity, road quality, and distance to site). Having the respondent choose the preferred option, the data from many such choices can be used to estimate the value of different options. With the advent of computerized survey methods, SCM has become more popular because the surveyor can offer many different options to the respondent using pictures or computer-generated images that differ in key environmental aspects (Thurston 2006; Dillman 2000).

All of the methods require significant data collection efforts because the estimates of WTP have to be as immune as possible to individual differences, and the generalization of the data is done through regression analysis (e.g., see the Glossary). That is what "maximum likelihood" is in the article quoted in Chapter 1. The choice of any of these methods depends on the circumstances and is artfully rather than scientifically chosen. If you feel, for example, that many people use or would use a stream if it were cleaned up for bass fishing and the area is otherwise a popular outdoor recreation area, it makes sense to look at the travel cost method. The general rule is to use revealed preference methods when you are primarily concerned with use values and the related markets exist or use stated preference methods when you are primarily concerned with nonuse values (and have the time and money to implement a quality study).

BENEFIT TRANSFER

Once the benefits and costs of a project are determined using one of the methods described, it is sometimes desirable to "transfer" those estimates to another site. A watershed group in Colorado might want to use the benefits estimated for restoration of a stream in Pennsylvania. Benefit transfer (BT) is appealing because it is not expensive, but there are many potential problems. What if the Colorado group knows that restoration will increase rainbow trout populations, but the Pennsylvania study found values for higher smallmouth bass populations? There also will be a difference in populations of people who are used to estimate, through stating or revealing their preferences, the benefits of an activity. One area may be richer, have a higher percentage of a given ethnic group, be generally younger, or otherwise have different demographic characteristics.

There are ways around this, however, and usually it involves recalculation of the values using the characteristics of the people at the new site, but without polling them directly; in that way, one saves time and money but gives up some specificity.

INPUT/OUTPUT ANALYSIS

Sometimes, information on the efficiency (i.e., do the benefits outweigh the costs?) of a project is not the type of support that the decision maker needs. One approach to address the economic impact of a decision or project is input/output (I/O) analysis. This analysis describes or predicts what the economy of a region, state, or even country is going to do on the addition of a new industry or economic activity. For example, if a new fast food store opens in the small town of Davis, West Virginia, the impacts will not be limited to the few people it employs. The employees will then have more money and will probably spend some of it in the town at other places of business, the owners of which will then have more money; this is the *multiplier effect*. In the case of stream restoration, I/O might be used to determine the negative impacts of closing a mining operation; folks would be out of work, and the multiplier would go in the opposite direction. Another example might be to measure the positive impacts of starting a restoration business in a community. Standard software is commercially available to do I/O analysis, but it requires data for large market areas (e.g., county boundaries).

EXAMPLES FROM THE LITERATURE

Now that you have a basic background on the methods, your next question might be, Where do I begin? We suggest examining economic studies that might be similar to the watershed issue, provided next. Although actual results are presented, they should only be considered as examples because the particular question or area may not match the environmental changes or the demographics of the original study. The first example describes a stated choice survey that examined people's preferences for restoring acid mine drainage streams in Pennsylvania. The second example describes an interesting contingent valuation study that examined WTP to remove an invasive species from Yellowstone Lake. A travel cost method study is summarized next; it focuses on activities like motor boating, camping, and sightseeing for different ecoregions. The next study uses the hedonic pricing method to examine how wetlands affect home prices. The final study described uses BT to value water quality changes in the Chesapeake Bay.

WATER QUALITY IN TWO PENNSYLVANIA WATERSHEDS: STATED CHOICE METHOD

In an article, "Valuing Watershed Quality Improvements Using Conjoint Analysis," Farber and Griner (2000) focused on two watersheds in Pennsylvania, the Loyalhanna Creek and the Conemaugh River, and estimated dollar values for incremental changes in water quality for both watersheds.

The authors used the SCM, sometimes called conjoint analysis, choice modeling, or contingent choice. The SCM is a method popular in marketing literature and enjoying a growing following among economists. Additional SCM studies applied to valuing environmental or natural resource goods include those of Milon and Scrogin (2006), who studied the value of restoring the Greater Everglades ecosystem; Adamowicz et al. (1997), who valued the characteristics of fishing sites; Hanley et al. (1998), who valued forest landscapes; and Rolfe et al. (2000), who examined loss of tropical rain forests.

In their study, Farber and Griner (2000) sent a survey to a total of 510 households (of 3,958 local residents) in summer 1996. People were presented with several different scenarios and were asked to choose between the status quo (think of this as the current condition or baseline condition) and various combinations of stream quality improvements for the two streams. Each alternative had a price attached. A total of 367 usable surveys were returned.

While this method has great appeal because it uses market-like choices, the surveys need to be carefully written to avoid being too long and too complicated. But, if results are required in a short amount of time, developing a new questionnaire may not be appropriate. To see how complex an SCM survey can become, consider the attributes and levels (or characteristics of the particular stream and changes to that stream) that describe acid mine drainage that go into the Farber and Griner survey's choice sets: Loyalhanna Creek changes (hypothetically) from moderately polluted to unpolluted, and the Conemaugh River can change from severely polluted to moderately polluted to unpolluted. Stream condition was expressed in the surveys based on survivability of fish and "other organisms." The article used five payment levels ($15, $45, $90, $180, $360) in the alternatives. So, the total number of potential alternative descriptions was 25, that is, $(2 \times 5) + (3 \times 5)$. However, there is also a status quo description, for a total of 26. Each alternative describes one possible scenario of stream restoration, and these alternatives are combined into a choice set. In this study, there were three alternatives, including the status quo. Each respondent was asked to respond to five choice sets (in Chapter 3, Appendix 3A, Collins et al. provide examples of their choice sets). When larger sets of attributes and levels are considered, software programs help simplify the choice sets so statistical conclusions are valid with a shorter survey. To separate out the use and nonuse values people place on stream restoration, Farber and Griner divided the survey respondents into users (households with members who had visited one of the streams within the past year) and nonusers. Socioeconomic characteristics such as distance to site and income were also incorporated into the model. Estimates of WTP varied depending on the present condition of the watershed and the amount of cleanup expected. For example, Farber and Griner estimated that households were willing to pay for 5 years (in 2005 $):

Conemaugh: Severe → Moderate: $45
Conemaugh: Severe → Unpolluted: $95
Loyalhanna: Moderate → Unpolluted: $34

You may ask yourself what (in 2005$) means. The Consumer Price Index (CPI) is a measure of the average change over time in the prices paid by urban consumers for a fixed market basket of goods and services. It is a standard measure of inflation. Therefore, we are trying to avoid comparing prices that are impacted by inflation. Economists like to compare prices and money using a base year (like 2005). You can find more information on the CPI at http://www.bls.gov/cpi.

Scale can be incorporated into the design of an SCM study by varying the size of the environmental good or by incorporating a broad set of levels. Developing separate questionnaires that describe different scales of the environmental good is possible but also costly and time consuming. To investigate the benefits of restorations at different scales would require more focus groups to understand how respondents perceive the good and how different regions perceive the good. If regions describe the good differently, multiple versions of the questionnaire may be required. Another possibility is to create a set of levels for each attribute that is broad enough to test for the most detailed description of the environmental good. However, it is difficult to incorporate all relevant characteristics or attributes in the design as scale increases. Respondents perceive more levels per attribute as the size of the environmental good increases. For example, if you are asking about one fishing site on one particular stream, then potential respondents are probably focused on such attributes as type of fish, travel time to get there, crowding, and expected catch rate. However, as you increase the scale, say to fishing sites in West Virginia, the levels of those attributes may have to increase to better represent people's thoughts, like both cold water and warm water fish species, types of scenery from flat to mountainous, boating versus wading versus bank fishing, and water quality. Finally, because economic analysis requires some type of change (e.g., a restoration option, global warming, etc.), respondents must perceive that the driver of change will affect all fishing sites in West Virginia. The policy has to match the scale of the natural resource in question.

Exotic Invaders in Yellowstone Lake: Contingent Valuation Method

A good recent example of the CVM comes from Alberini and Kahn's (2006) *Handbook on Contingent Valuation*. This book is a good reference and divides the edited chapters in it into three sections: one section on economic theory of CVM, one on econometric issues (see Glossary), and one with several case studies. One of the case studies is by Cherry et al. (2006), "Valuing Wildlife at Risk from Exotic

Invaders in Yellowstone Lake." The authors noted that starting in the mid-1990s people were catching lake trout in Yellowstone Lake. Lake trout are an invasive species, probably first transported to Yellowstone by a visitor, that have since started taking over. Because they swim and spawn in the deep waters of the lake, lake trout displace the native cutthroat trout without really replacing them in the food chain. Cutthroats spend more time in the shallows, becoming food for osprey, white pelicans, and other birds of prey. Furthermore, cutthroats swim upstream to spawn in the lake's tributary creeks, where many end up as food for grizzly bears. The reduction in population, or even possible decimation of the cutthroat, by the competing lake trout would, according to ecologists, have a noticeable impact on the number of sightings of the fish's natural predators. As wildlife viewing is, as the authors pointed out, an extremely important activity for most of the park's visitors, lowered probability of sighting these megavertebrates has a potential economic as well as ecological impact on the park.

Control of the lake trout population is an expensive proposition. The park has budgeted about $250,000 a year to deep net the trout. So, one might wonder if this is a cost-effective use of the park's limited funds. The authors (Cherry et al. 2006) used CVM to estimate the park visitors' WTP for such a policy. The authors handed out surveys to 496 people over the course of 3 days at one of the entrances to the park. They asked people to return the surveys within 3 weeks; 284 were returned. The response rate was, as the authors noted, quite high, 57.3%. The authors attributed that to the fact that those who did not want to participate simply refused to take a blank survey when offered; also, the people the researchers approached were people who obviously had an interest in what was going on in Yellowstone Park. The authors also noted that they were not able to follow up with people who were handed surveys, and this would decrease the response rate.

Cherry et al. (2006) used a two-step method for analyzing the data from their survey. They allowed for zero WTP by asking first if the respondent would be willing to contribute anything to the hypothetical "Yellowstone Lake Preservation Fund." If the respondent answered "yes," the respondent was given a choice among three randomly assigned amounts ($5, $15, and $30) and was asked if he or she would agree to pay that amount yearly to the fund for the purpose of controlling the lake trout population. The authors also collected demographic data from the respondents and on the survey asked some questions about their familiarity with the particular invasive species problem addressed.

The authors (Cherry et al. 2006) estimated through this survey that the average park visitor was willing to pay approximately $11 per year to fund a program designed to control the lake trout population. Dividing the $250,000 per year that the current program costs over the estimated 3 million annual visitors clearly is less than $11, and the authors concluded: "Our results indicate that visitor benefits clearly outweigh the cost of current policy."

Cherry et al. (2006) also mentioned some caveats that even a well-designed survey like theirs had to take into account. First, since people are focusing on one species when taking the survey, valuing species in a piecemeal fashion like this is likely to give an overestimate when CVM results for multiple species are added together. Second, especially when endangered species are included in the mix, a CVM survey "provides respondents with a chance to state their general preferences toward the entire gamut of endangered species, not just for the specific species in question" (p. 320). Nevertheless, Cherry et al., through the use of CVM, provided a convincing and transparent argument for continued support of a program to protect the cutthroat trout in Yellowstone Lake even though the up-front costs are relatively high.

ISSUES OF SCALE

After the Valdez oil spill, the U.S. National Oceanic and Atmospheric Administration (NOAA) convened a "blue ribbon" panel of economists to assess the use of CVM to measure environmental damage (specifically, nonuse values). The panel's qualified endorsement of the method appeared in the work of Arrow et al. (1993). In the article, the authors set forth a series of recommendations; one of the most important ones was that the CVM study pass a test of scope. *Scope* (also known as embedding, part-whole or nesting) problems occur when survey respondents answer that one aspect of a system they are being asked to value has a value that is relatively or even absolutely higher than the whole system. For example, suppose one group of survey respondents is asked to give its WTP for preservation of elephants in the Serengeti, and another similar group is asked a question that leads to valuation of the Serengeti itself (including elephants, other species, and the entire habitat); if the data estimate similar values, then we probably have scope effects.

The test for scope effects requires that a split-sample data set be collected, and that rigorous statistical analyses be done. Berrens et al. (2000) dealt with embedding in their research valuing in-stream flows and silvery minnow in the middle Rio Grande. The silvery minnow is an endangered species with habitat that occurs uniquely in the Rio Grande in the stretch between Cochiti Dam and the Elephant Butte Reservoir. Berrens et al. carried out a CVM survey intended to estimate the WTP for habitat preserving in stream flows in the middle Rio Grand valley. In addition to asking a standard WTP question for the change in in-stream flows, the authors asked a subgroup of 561 respondents a question that was "modified to identify minimum in stream flows to specifically protect the silvery minnow" (p. 79). The estimated coefficients and WTP were significantly different and lower than those estimated from the group of 564 whose WTP question did not specify the silvery minnow. So, they did not have the problem. While one cannot eliminate all bias in any survey-based valuation method, obvious sources must be identified and minimized when possible and explicitly mentioned when not.

OUTDOOR RECREATION: TRAVEL COST METHOD

A good article describing the travel cost method is that of Bhat et al. (1998). These authors described an individual travel cost method to estimate the economic value of outdoor recreation for different ecoregions. The activities they examined were motor boating and waterskiing, developed and primitive camping, cold-water fishing, sightseeing and pleasure driving, and big game hunting. The ecoregions they focused on are the Northeast and Great Lakes (Wisconsin, Illinois, Indiana, Ohio, Michigan, Kentucky,

TABLE 2.1
Bhat et al. (1998) Consumer's Surplus Per Day by Activity Across Ecoregions

	Motor Boating and Waterskiing	Camping	Cold-Water Fishing	Sightseeing and Pleasure Driving	Big Game Hunting
Northeast and	9.85	261.12[a]	N/A	13.90	4.31
Great Lakes	(9.12, 10.58)			(13.12, 14.68)	(3.92, 4.70)
New England	N/A	43.25	N/A	N/A	N/A
and warm		(22.88, 63.62)			
continental					
Appalachian	43.61	6.39	25.70	23.73	6.09
Mountains	(35.70, 51.52)	(5.58, 7.20)	(20.47, 30.92)	(21.88, 25.58)	(4.89, 7.29)

N/A, not applicable because of data limitations.

[a] Based on insignificant price coefficient (i.e., the particular variables were found not to be important in the model).

Tennessee, West Virginia, and Pennsylvania); Appalachian Mountains (Maryland, West Virginia, Virginia, North Carolina, South Carolina, Tennessee, Georgia, and Alabama); and New England and warm continental (Maine, New Hampshire, Vermont, New York, Connecticut, Rhode Island, New Jersey, and Pennsylvania).

This method was originally developed for individual sites, but Bhat et al. (1998) used it to estimate demand functions at a larger scale (i.e., various activities within an ecoregion). Therefore, the method requires more aggregate information about the activities than with individual sites. They collected information on annual trips by individuals to different ecoregions for various activities; they got information on household income levels, cost of the trips, and costs of logical substitute trips.

Using data from the Public Area Recreation Visitors Study and their survey, Bhat and coworkers (1998) estimated individual demand curves. Based on the results of their statistical model, they estimated value per day in each ecoregion. The value (or consumer surplus) represents the change in an average individual's welfare (or well-being) by increasing outdoor recreation days. For example, a policy would have to increase the number of outdoor recreation days for an activity across the ecoregions. Table 2.1 redisplays the mean net economic value across the ecoregion for activities.

The method faces many limitations (e.g., what to do about individuals who travel to multiple sites or data requirements). If decision makers are only interested in ecoregional values of outdoor recreation or even values of outdoor recreation for a specific site, then this method would be useful.

Wetlands in North Carolina: Hedonic Pricing Method

A study by Bin and Polasky (2005) used the hedonic pricing method to calculate the effects of wetlands on rural North Carolina home prices. They used parcel data from Carteret County, wetland data from the North Carolina Division of Coastal

Management, flood maps from the Federal Emergency Management Agency, and demographic data from the 2000 U.S. Census. All data were connected using a geographic information systems (GIS) so that properties have surrounding wetlands connected to it. Their model is able to distinguish changes in property values based on increases in the size of the closest wetland, decreasing the distance to the closest wetland, and increasing the acreage of wetlands within ¼ mile of the property. All of these components reduce the rural property values, which differs from previous hedonic studies for urban areas. For example, Bin and Polasky stated that property values drop by approximately $350 to $2,300 if the distance decreases by 10% to 50% (evaluated at the mean distance of 766 feet). Increasing the size of the nearest wetland by 10% to 50% reduces property value but not as much as distance ($93 to $394, evaluated at the mean size of approximately 57 acres). Some may have trouble understanding why wetlands, which do a lot of good for the environment, would reduce property values. Wetlands in urban areas are found to increase property values. Values depend on people's perceptions of the particular change in the service or good. If an individual thinks a wetland would prevent a particular land use (e.g., you cannot fill a wetland in without a permit), the value of that wetland may be lower. The authors suggested that if wetlands are in short supply, like in urban areas, then the value may be higher compared to when wetlands are abundant (like in this particular rural county). The value of a particular good or service decreases as more and more of it is provided. We refer to this as *decreasing marginal benefits,* and it is not unique to this study. If house prices are available for a particular watershed area, this method may be appropriate, but the key is finding environmental quality data that homeowners will perceive. In addition, this method will not get at visitors' benefits, so watershed groups need to consider the importance of this in their particular decision.

COMBINING STUDIES: BENEFIT TRANSFER

A low-cost and increasingly common method for integrating economic values into ecological policy choices is the BT method. Valuation is done by applying data that are collected elsewhere to the study site to inform a decision. BT is the practice of adapting value estimates of a quality or quantity change for some environmental resource to evaluate a proposed change in a similar resource.

Practitioners usually transfer either the *benefit value* or the *benefit function.* To use the benefit value, one adds up the average values estimated elsewhere. The benefit function is used by incorporating the appropriate variables into the original equation. An obvious drawback of either is the introduction of secondary data, while the primary benefits of the method are that it saves time and money and, if applied judiciously, gives credible, useful figures. Smith et al. (2000) made several recommendations to ensure sound analysis using BT. For a thorough overview of the method itself and a detailed analysis of underlying theory and some of the method's common pitfalls, the reader is encouraged to consult Volume 28 of *Water Resources Research* (1992), a special issue dedicated to research on BT.

Morgan and Owens (2001) illustrated the application of the method to measure the benefits of water quality in the Chesapeake Bay. The authors identified six major

TABLE 2.2
Morgan and Owens (2001) Benefit Estimates[a] in Million Dollars

	Percent Improvement								
	20%			40%			60%		
Activity	Low	Average	High	Low	Average	High	Low	Average	High
Beach use	15.1	48.0	62.3	36.2	55.6	89.9	288.8	824.9	1,520
Trailered boating	0.90	6.5	11.2	1.36	9.97	17.2	6.7	48.5	83.6
Striped bass fishing	0.92	1.89	2.86	5.13	10.3	15.4	62.4	128.6	194.8
Total	16.9	56.4	76.4	42.69	75.87	122.5	357.9	1,000	1,800

[a] All estimates in 1996 dollars (see Chapter 4 for a discussion of converting dollars to the same base year).

categories of relevant benefits: recreation, fishing, health, property values, regional economic impacts, and nonuse values. The authors noted that loading of phosphorus and nitrogen is a leading cause of degradation in the Chesapeake Bay. Using data from studies published in other articles on the Chesapeake Bay (Bockstael et al. 1988, 1989; Krupnick 1988), the authors valued the difference in water quality in 1996 to the quality it would have been in 1996 had the Clean Water Act (CWA) not been enabled. Since the CWA controls nitrogen and phosphorus in waterways, it serves as the policy change of interest in this study. Remember, we mentioned that when conducting any economic valuation analysis, we always need to do it in the framework of some kind of change to the environment.

For a 20%, 40%, and 60% increase in water quality, Morgan and Owens (2001) estimated low, average, and high dollar values of the increase in water quality it is predicted the CWA will affect. The results are given in Table 2.2.

Smith et al. (2000) noted that BT involves four general steps:

1. Translate the policy change into one or more resulting quantity changes for uses that are linked to an environmental resource.
2. Estimate the number of typical users before and after the policy change.
3. Transfer a per "unit" consumer surplus measure, with the unit measure comparable to the index used in step 1.
4. Combine estimates in steps 1 through 3 for each year considered in the analysis and compute the discounted aggregate benefit measures.

One can see that scale issues will come into play primarily in steps 1 and 2. In step 1, ecological scale effects must be taken into consideration, and in step 2 the analyst needs to determine the number of users (obviously a scale issue) and define the term *typical*, which may change as the scale of the project goes from local to regional to larger. The usefulness of BT depends a lot on how specific one needs the figures to be and the availability of other studies that have already collected data like that needed in the current area of interest.

MORE ON SCALE

While economic valuation of ecological or nonmarket goods is a common practice, there is a tight set of circumstances and assumptions under which it can be done accurately. Among these are the existence of a well-defined policy goal, an intimate knowledge of the decision makers and stakeholders affected, and a scientific understanding of the ecological changes that the means to the policy goal will affect. One important thing you might notice in most of this chapter is that we have tried to point out when issues of scale are important and how dealing with scale is interesting and hard. If we were able to change the scale of the analysis simply by multiplying the area of interest by the appropriate constant, the solution would be easy. But, it is not like that. As you scale up, you often change the makeup of the thing you are trying to evaluate. Sometimes, it just goes up smoothly, but sometimes when you talk about a larger scale you add some entirely new aspect to the problem. Think about a watershed, for example. There is a big difference, not just in the area it takes up, between a small subwatershed of headwater and second-order streams, and a large watershed (i.e., five-digit hydrologic unit code [HUC]) with different water currents, different habitat, different critters, different opportunities for recreation, and so on.

SUMMARY

When reading the rest of the chapters, look for clues regarding why the authors of the study chose a certain kind of valuation method: Did they use a stated or revealed preference method and why? What circumstances were present that allowed them to do a travel cost or a hedonic study? If the authors do not talk about how the results were used, can you think of what decision tool you might have used to formally present the findings?

REFERENCES

Adamowicz, W., J. Swait, P. Boxall, and J. Louviere. 1997. Perceptions versus objective measures of environmental quality in combined revealed and stated preference models of environmental valuation. *Journal of Environmental Economics and Management* 32:65–84.

Alberini, A., and J. Kahn. 2006. *A Handbook on Contingent Valuation.* Northhampton, MA: Elgar.

Arrow, K., R. Solow, P. Portney, E. Leamer, R. Radner, and H. Schuman. 1993. Report of the NOAA Panel on Contingent Valuation. *Federal Register* 58(10): 4602–4614, January 15.

Berrens, R., A. Bohara, C. Silva, M. McKee, and D. Brookshire. 2000. Contingent valuation of instream flows in New Mexico: With tests of scope, group-size reminder and temporal reliability. *Journal of Environmental Management* 58(1): 73–90.

Bhat, G., J. Bergstrom, R.J. Teasley, J. Bowker, and J.K. Cordell. 1998. An ecoregional approach to the economic valuation of land- and water-based recreation in the United States. *Environmental Management* 22(1): 69–77.

Bin, O., and S. Polasky. 2005. Evidence on the amenity value of wetlands in a rural setting. *Journal of Agricultural and Applied Economics* 37(3): 589–602.

Bockstael, N.E., K.E. McConnell, and I.E. Strand. 1988. *Benefits from Improvements in Chesapeake Bay Water Quality*. Environmental Protection Agency Cooperative Agreement CR-811043-01-0.

Bockstael, N.E., K.E. McConnell, and I.E. Strand. 1989. Measuring the benefits of improvements in water quality: The Chesapeake Bay. *Marine Resource Economics* 6(1): 1–18.

Cherry, T.L., J. Shogren, P. Frykblom, and J. List. 2006. Valuing wildlife at risk from exotic invaders in Yellowstone Lake, in *A Handbook on Contingent Valuation*, A. Alberini and J. R. Kahn, Eds. Northampton, MA: Elgar.

Dillman, D.A. 2000. *Mail and Internet Surveys. The Tailored Design Method*, 2nd edition. New York: Wiley.

Farber, S., and B. Griner. 2000. Valuing watershed quality improvements using conjoint analysis. *Ecological Economics* 34:63–76.

Hanley, N., R. Wright, and V. Adamowicz. 1998. Using choice experiments to value the environment. *Environmental and Resource Economics* 11(3–4): 413–428.

Krupnick, A. 1988. Reducing bay nutrients: An economic perspective. *Maryland Law Review* 47(2): 453–480.

Milon, J.W., and D. Scrogin. 2006. Latent preferences and valuation of wetland ecosystem restoration. *Ecological Economics* 56(2): 162–175.

Minnesota IMPLAN Group, Inc. *IMPLAN Professional*. Version 2.0: User's Guide, Analysis Guide, and Data Guide. June 2000.

Morgan, C., and N. Owens. 2001. Benefits of water quality policies: The Chesapeake Bay. *Ecological Economics* 39:271–284.

Rolfe, J., J. Bennett, and J. Louviere. 2000. Choice modelling and its potential application to tropical rainforest preservation. *Ecological Economics* 35(2): 289–302.

Smith, V., G. Van Houtven, S. Pattanayak, and T. Bingham. 2000. *Improving the Practice of Benefit Transfer: A Preference Calibration Approach*. Washington, DC: U.S. Environmental Protection Agency, Office of Water.

Thurston, H. 2006. Non-market valuation on the Internet, in *A Handbook on Contingent Valuation*, A. Alberini and J.R. Kahn, Eds. Northampton, MA: Elgar.

3 Valuing the Restoration of Acidic Streams in the Appalachian Region
A Stated Choice Method

Alan R. Collins, Randall S. Rosenberger,
and Jerald J. Fletcher

CONTENTS

INTRODUCTION

Acidification is a major water quality problem in the Appalachian region of the United States. Of the over 5,000 stream miles that are impacted by acidification in this region, about 90% is a result of coal mining (U.S. Environmental Protection Agency [USEPA] 2002). This impact is commonly referred to as *acid mine drainage* or AMD. Problems associated with AMD include the contamination of public drinking water and industrial water supplies, poor growth and reproduction of aquatic plants and animals, reduction in recreational fish species, restricted stream use for recreation, and corroding effects on bridges.

Given the need for restoration of streams affected by AMD, state and federal agency officials have been struggling with issues of how to

1. Justify stream restoration within a benefit-cost framework
2. Prioritize restoration projects among the numerous degraded streams given limited budgets
3. Demonstrate the economic importance of preserving stream quality where degradation has not occurred
4. Devise a cost-efficient method of data collection for economic valuations

These concerns were expressed by representatives from the Natural Resource Conservation Service (NRCS), the West Virginia Soil Conservation Agency, the West Virginia Division of Environmental Protection (WV-DEP), the Canaan Valley Institute, and the Rivers Coalition at a stream valuation workshop held October 2000 at West Virginia University. Since minimal research has been conducted on the valuation of stream restoration (Farber and Griner 2000), this study was undertaken to provide important information by designing and testing a method for valuation of stream restoration that uses a combination of Internet and mail surveys.

The objectives of this study were to

• Create a survey device that allows for effective data collection of economic value information within an AMD-impacted watershed
• Determine economic values for different levels of stream restoration using a stated choice method (SCM) approach

Much like the contingent valuation method, the SCM uses surveys based on the simple idea that if an analyst wants to know people's maximum willingness to pay (WTP) for an environmental good or service (like stream restoration), you simply ask them via a constructed or hypothetical market (Champ et al. 2003). Maximum WTP is the highest price that a survey respondent would pay to obtain the environmental good or service.

Frequently, stated, as opposed to revealed, preference methods are the only methods available for estimating monetary values for nonuse goods. Since individuals are asked to declare their preferences via hypothetical payments, no market actions are observed with this survey method. SCM has proven useful in decision making, particularly regarding legal damage awards from environmental degradation (Mitchell and Carson 1989; National Research Council 2005).

DESCRIPTION OF THE STUDY AREA

Deckers Creek watershed is located in Monongalia and Preston Counties of West Virginia (Figure 3.1). This watershed contains an area of about 63 square miles. Deckers Creek flows 23.7 miles from southeastern Monongalia County, into Preston County, and then back into Monongalia County before emptying into the Monongahela River at Morgantown, West Virginia. There are three distinct portions of Deckers Creek:

1. Beginning as a small woodland brook, it flows through a long, flat valley as a straightened ditch among agricultural fields above Masontown, West Virginia.
2. A white water middle portion starts below Masontown and cuts a path through a narrow gorge that contains limestone outcroppings.
3. A flat portion begins at Dellslow, West Virginia, and ends at the Monongahela River. This third portion is heavily impacted by AMD, particularly from an abandoned underground mine discharge (Figure 3.2).

Deckers Creek has a number of contamination problems that are typical of rural Appalachia—trash in the creek, sewage, and AMD contamination. WV-DEP has listed the entire length of Deckers Creek on their 303(d) list of impaired streams. While stream water pH levels have been increasing slowly over the years (Stewart and Skousen 2003), acidic conditions still inhibit most aquatic life throughout the lower portion of the creek. In addition, there are elevated levels of sulfates, iron, aluminum, and manganese. Since Deckers Creek is not used as a drinking water supply, these contaminants do not present hazards to human health. A $10 million restoration plan has been drafted by state and federal agencies, but funding is only now being secured by the Friends of Deckers Creek (FODC) to complete this restoration.

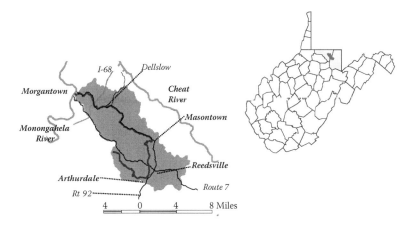

FIGURE 3.1 Map of Deckers Creek Watershed and its location within the state of West Virginia.

(a)

FIGURE 3.2   (b)

(c)

FIGURE 3.2   *(continued)* Deckers Creek (a) near Zinn Chapel, June 2002; (b) Route 7 above Pioneer Rocks, December 2001; (c) behind Tramps Bar, August 2002.

Local interest in restoring Deckers Creek is high. FODC is a very active watershed association dedicated to restoration of the stream. A rail trail along the creek provides recreational access to the creek and creates a high level of awareness about the creek among local citizens using the rail trail. Thus, stream restoration could have significant impacts on direct use of the stream (fishing, kayaking, etc.) as well as indirect effects on the value of rail-trail recreational experiences.

RESEARCH METHODS

SURVEY

Given the relatively small size of Deckers Creek, the populations most impacted by its restoration were assumed to be people living within the watershed and users of the creek and rail trail. Surveys of two potentially different populations in the Deckers Creek watershed were used to collect data: (1) the general public and (2) stream users. Thinking about the kinds of values mentioned in Chapter 1, the general public would include people who have both use and nonuse values, while the stream users would probably focus on use values. Prior to conducting the survey, citizen attitudes and values about Deckers Creek restoration were determined using focus groups held during the fall of 2001.

Collins et al. (this chapter) bring up a good point about survey research. Before any survey can be sent out, researchers need to know how the community talks and feels about a particular environmental good or service. Focus groups are one part of developing the questionnaire. Other approaches include one-on-one interviews, sometimes referred to as verbal protocols. Verbal protocols, a technique that has respondents "think aloud," can be used to determine what respondents are thinking about as they read and answer the questionnaire. Pretesting, when a small sample is used to test the questionnaire, is also important. Information about respondents rejecting a question and the potential response rate can be estimated from this activity.

Three focus groups ranging in size from 8 to 15 people were conducted with local citizens and members of FODC. For some good guidelines on how to conduct a focus group, see the book by Krueger (1994). From these focus groups, three important attributes of stream restoration on Deckers Creek were identified: aquatic life, scenic quality, and swimming/wading. There are linkages between restorations of each attribute. For example, correction of AMD problems would restore aquatic life and would improve some aspects of swimming/wading. However, restoration of one attribute in Deckers Creek does not necessarily improve the other two. Correction of AMD restores aquatic life but would not eliminate trash problems (scenic quality), and increasing the pH actually makes bacteriological problems from sewage worse for swimming/wading as a low pH inhibits bacteria growth.

Following the focus groups, electronic and paper copy survey instruments were developed and tested with FODC members, the general public, and students at West Virginia University. Design of the electronic survey followed recommendations from Dillman (2000). Survey questions included respondent recreation behavior related to public waterways and parks, attitudes about stream restoration in general, knowledge about Deckers Creek, restoration choices, and the usual demographic characteristics. The electronic survey was made available to access code holders via an Internet Web site. See Appendix 3A for a copy of the mail survey used in this study.

Four restoration choice questions were presented to each respondent. Each choice question included three options describing three stream quality attributes (aquatic, scenic, and swimming) and a cost attribute (represented as an increase in monthly utility bills). Based on focus group responses and the current conditions of Deckers Creek, a status quo option was provided in each choice question. This status quo option represents the current conditions of the stream where all three stream quality attributes were at low quality levels and a zero additional cost for monthly utility bills.

In the other two options, stream restoration attributes were randomly assigned two levels—moderate or complete (referred to as "High" in the survey). See Appendix 3B for descriptions of each attribute and restoration levels. We needed to include a status quo option because, from the discussion in Chapter 1, our nonmarket valuation techniques rely on asking people to value a small change in the resource.

The survey was conducted from October 2002 to August 2003. Within the watershed, people were contacted via telephone and asked to participate in either a mail or an electronic survey. A random sample of residential telephone numbers was obtained from Survey Sampling Inc. Calling was done by five West Virginia University students during October and November of 2002 and then in February and March of 2003. At least three attempts were made to contact each phone number. Once respondents agreed to participate, either a paper copy of the survey was mailed to them or they were e-mailed the Web site address of the electronic survey. In a few cases, the Web address was sent to them through the mail, or the Web address and appropriate access code were given to them over the phone.

A random sample is important for a survey because it is one step in getting a representative response to your questions. What you do not want to happen is to gather data from one particular group in your population. That is, if you want to know the values that those who fish in West Virginia have for fishing sites, you would not want to just send out surveys to those who fish in cold water. Warm-water anglers have different opinions that could be important to your survey. Therefore, you want anyone in your population of interest to have an equal chance of answering your questionnaire.

Additional survey data were obtained from surveys conducted with two groups of creek and rail-trail users: recreational users of the rail trail along Deckers Creek and citizens committed to watershed improvements (FODC members). Throughout July and August 2003, personal interviews of rail-trail users were conducted. This survey was conducted at two locations along the rail trail. For FODC members, a solicitation e-mail was sent during May 2003 asking them to participate in the survey. Interested respondents who replied had the Internet survey information sent to them via e-mail.

ESTIMATION PROCEDURES

Once collected, survey data from an SCM question are used to estimate a maximum WTP for an environmental improvement among the sample respondents. This maximum WTP is then aggregated to the entire population. Some SCM studies require relatively uncomplicated procedures to estimate maximum WTP for environmental improvements. When respondents are asked directly about their maximum WTP, then computing a simple average of maximum WTP values can be done. Aggregation also is simplified when the sample is assumed to be similar to the population. In this case, the average maximum WTP from the sample was multiplied by the number of households in the population.

In the Deckers Creek study, however, respondents were not asked directly about their maximum WTP for stream restoration. Rather, respondents were asked a series of four choice questions using multiattribute, choice experiments (Louviere et al. 2000). These four questions each included three options, of which respondents were to choose one (restoration option A, restoration option B, or status quo option [no restoration]). Each restoration option was composed of randomly assigned levels for each of the stream quality attributes plus a cost attribute. Since we needed to use the choice information along with the three stream quality attributes and a cost attribute presented to each respondent, we could not use the easy estimation methods and had to use what is called a "two-level nested logit" model to estimate maximum WTP (Collins et al. 2005).

The term logit refers to the kind of statistical estimating technique used. The choice of technique is best left up to a statistician or econometrician, but there are several techniques to choose from depending on how you think your data are distributed. These include variants on the logit and probit methods.

A two-level nested logit model has a number of advantages: (1) maximum WTP differences between sample groups can be estimated; (2) maximum WTP values can be determined for restoration of the different stream attributes (aquatic vs. scenic vs. swimming) or for different levels of restoration within an attribute; and (3) aggregate total monetary values for different combinations of attributes can be computed.

They call these models *nested* because there is a decision based on, or nested in, another decision. The appeal of this type of model is that it represents how people sometimes make decisions. For example, a person might first decide what to do for recreation (e.g., go fishing) and then decide where to go fishing (which lake or stream); the first decision is nested in the second. In our application, we assume respondents first decide whether to support restoration, and if restoration is chosen, then which one of two options they prefer most. One drawback of our application is that maximum WTP estimates are based on an improvement from moderate to complete restoration rather than from status quo to complete restoration. Another possible different experimental design might allow estimation for any level from no restoration to partial restoration to complete restoration for any one or all of the attributes.

To compute the economic value of stream restoration, we aggregated up from the sample data to the watershed population using a conservative approach for full (complete restoration of all three stream attributes) and partial (complete restoration of one stream attribute at a time) restoration. The following assumptions were used in a conservative approach:

- Separate monthly household maximum WTP estimates were used for those who fished and for those who did not. The population percentages of those who fished (38%) versus those who did not (62%) in the watershed were estimated from survey data of the general population.
- Those respondents who declined to respond to the survey were assigned a zero WTP value from restoration. Based on the number of survey responses divided by surveys sent out plus "no" responses over the phone, maximum WTP estimates on a per household basis were applied to 35.4% of watershed households of those who fished and those who did not. The maximum WTP estimates were adjusted downward to account for the no restoration choices among respondent households (an 8.5% reduction for those who did not fish and a 13% reduction for those who did).
- The total number of households in the watershed (35,719) was based on data from seven zip code areas that overlap parts of the watershed.

RESEARCH RESULTS

Survey

For the watershed population, a total of 1,716 phone numbers were called, of which 1,371 were residential numbers. A sample of 584 households completed the telephone portion of the survey. A total of 387 respondents agreed over the phone to complete a survey, either by mail or by the Internet. The overall response rate for completed stream valuation surveys was 53%, slightly higher for mail surveys (55%) compared to Internet surveys (51%). Considering the completed surveys received divided by the total telephone contacts, an overall response rate of 15% was achieved. While this response rate is on the low side for surveys, telephone contact was found to achieve a slightly higher response rate at a lower cost per completed survey than using mail contacts to solicit survey participation (Street 2005).

A total of 50 rail-trail users and members of FODC responded to the survey. The initial plan was for all of these surveys to be completed electronically. However, all rail-trail users ultimately completed paper copies due to difficulties in using a laptop computer along the trail. Of the 11 FODC members, only two completed the surveys electronically. The other nine respondents completed paper copies due to the server support system for the survey being "hacked" and down for a couple of weeks.

With the exception of education, both groups of respondents had similar characteristics with respect to the watershed population (Table 3.1).

There were differences between the two respondent groups: general population versus users. The majority of the general population was female (60%) compared to only 40% of the user sample. Major differences in age were found as 73% of

TABLE 3.1
Demographics of the Survey Subsamples and the Watershed Population

Demographic Characteristics	General Population Sample (*n* = 207)	User Sample (*n* = 50)	All Respondents (*n* = 257)	2000 Census Data from the Watershed[a]
Gender				
Female	60%	40%	56%	50%
Age: distribution of adult population				
18 to 45	47%	73%	52%	62%
46 and over	53%	27%	48%	38%
Education				
College degree	53%	67%	56%	36%
Household income:				
average	$43,000	$44,000	$43,000	$41,000

[a] Based on a population-weighted average of census data from zip codes located in the Deckers Creek watershed.

the users were 45 years or younger compared to 47% of the general population. Education attainment was higher among the user group compared to the general population (67% vs. 53% with a college degree), and both were much higher than that in the watershed (36%). Incomewise, however, both groups had similar household averages, between $43,000 and $44,000 annually. This average was very close to the 2000 U.S. Census average of $41,000 for the study area.

Responses to knowledge and attitude questions about stream restoration are presented in Table 3.2. The vast majority of respondents (77%) were familiar with at least the lower portion of Deckers Creek (Table 3.2). Overall, relatively few users (13%) were completely unfamiliar with Deckers Creek. Three-fourths of all respondents stated that there were environmental problems with Deckers Creek. Very few respondents (3%) thought that there were no environmental problems with Deckers Creek, although 22% of respondents stated they did not know of any environmental problems associated with Deckers Creek. As expected, the user group was more familiar with Deckers Creek environmental problems. Respondents stated the top three environmental problems associated with Deckers Creek were trash, unnatural colors, and lack of aquatic life (Table 3.2). Respondents perceived that the most widespread stream pollution problems in West Virginia streams were related to visual aspects (trash followed by acid and minerals) rather than mainly water quality degradation from sewage.

MAXIMUM WTP ESTIMATES

Maximum WTP was computed with data combined from both groups, users and the general population, because we found no statistical difference in the estimated WTP

TABLE 3.2
Respondent Knowledge about Deckers Creek and West Virginia Stream Water Quality

Question	General Population Sample ($n = 207$)	User Sample ($n = 50$)	All Respondents ($n = 257$)
What portion(s) of Deckers Creek are you familiar with?			
Lower portion	75%	83%	77%
Middle portion	44%	54%	46%
Upper portion	19%	33%	21%
I am not familiar with any portion	14%	13%	13%
Do you think there are environmental problems with Deckers Creek?			
Yes	73%	84%	75%
No	3%	2%	3%
Don't know	23%	14%	22%
What do you think are the main environmental problems with Deckers Creek?			
Unnatural colors	71%	77%	72%
Odor	54%	58%	55%
Lack of aquatic life	69%	77%	71%
Trash	84%	79%	83%
Unsafe to swim	56%	51%	55%
Unsightly development	39%	40%	39%
High levels of acid	66%	72%	67%
Very widespread pollution problems in WV streams			
Sewage	26%	35%	28%
Acid and minerals	43%	39%	42%
Trash	44%	43%	44%

between the two groups. For partial restoration, the aquatic attribute had the largest WTP, with scenic and swimming having roughly the same, lower WTP (Table 3.3).

When respondents were those who fished, the maximum WTP for restoration of aquatic habitat was more than doubled (from $5.09 to $12.16 per month increase), but swimming quality restoration was essentially reduced to zero (from $3.55 to $0.21 per month).

When all three stream attributes were completely restored (full restoration), maximum WTP was estimated to be 33% higher for those who fished than for those who did not ($16.09 vs. $12.37 per month). These maximum WTPs were interpreted to mean that respondents perceived restoration of Deckers Creek to be much more valuable when completely restored as compared to a moderate level of restoration. For example, the aquatic attribute improved respondents' valuation dramatically when the stream resource could be restored to a self-sustaining aquatic habitat compared

TABLE 3.3
Maximum WTP Estimates for Partial (by Stream Attribute)
and for Full Restoration (All Three Stream Attributes) of
Deckers Creek by Sample Respondents

	Maximum WTP Estimates ($/household/month)
Partial restoration	
Aquatic	$5.09
Scenic	$3.72
Swimming	$3.56
Aquatic (for fishermen)	$12.16
Swim (for fishermen)	$0.21
Full restoration	
Fishermen	$16.09
Non-fishermen	$12.37

to restoration that was dependent on fish-stocking programs (as a moderate level of restoration would achieve).

Table 3.4 shows the maximum WTP estimates aggregated up to the entire Deckers Creek watershed population. This estimate was interpreted as the annual benefit that the watershed population would gain when the creek is restored. For full restoration of all three attributes, an estimated $1.9 million in restoration benefits occurs annually. Restoring aquatic life in Deckers Creek was responsible for most (56%) of this monetary benefit. These benefits represent conservative estimates as only about one-third of households in the watershed were estimated to place a positive economic value on restoration.

TABLE 3.4
Maximum WTP Estimates on an Annual Basis Aggregated over the Deckers
Creek Watershed Population

	Partial Restoration			Full Restoration
	Aquatic	Scenic	Swimming	Aquatic, Scenic, Swimming
Annual maximum WTP	$1,049,000	$507,000	$317,000	$1,873,000
Percent of full restoration	56%	27%	17%	

CONCLUSIONS FROM THIS STUDY AND ISSUES WHEN USING THE STATED CHOICE METHOD

The SCM utilized in this study allowed for assessment of maximum WTP when restoring individual attributes of Deckers Creek—aquatic habitat, swimming, and scenic quality. Maximum WTP estimates for improvements from moderate to complete restoration of all three stream attributes ranged between $12 and $16 per month per household. Respondents who fished had the largest WTP for restoration, and the majority of WTP among all respondents was for improvement in aquatic habitat. These WTP estimates were regarded as reasonable given that they represent about 25% to 35% of the average water and sewer utility bills for a Morgantown household in Monongalia County. When maximum WTP estimates were aggregated up to the entire watershed population, the estimated benefit from restoration of Deckers Creek was about $1.9 million annually. This benefit estimate probably underestimates the entire gain from restoration because it does not include valuation of stream improvements that may be derived from partially restoring each attribute of Deckers Creek to a moderate level of restoration.

Aggregated maximum WTP estimates can be used to assess the monetary benefits of restoring Deckers Creek. These monetary benefit estimates provide additional motivation for implementation of restoration efforts and incorporate public desires for restoration into decision making. When these monetary benefits are compared to the monetary costs of restoration, a favorable benefit-to-cost ratio (i.e., greater than 1.0) for restoration can assist in shifting federal and state government funding priorities to Deckers Creek.

There are some issues for watershed managers to be aware of when using SCM to assess the value of restoration. These issues arose in this study and could well arise in other valuations of watershed restoration. They include the following:

- If SCM is utilized as a valuation approach, the outcomes of restoration described in the SCM question may not always correspond exactly to actual outcomes. Restoration is a complex undertaking with uncertain time frames and final outcomes. To be understandable to the general public, SCM questions typically do not incorporate all these elements of uncertainty (National Research Council 2005).
- How do maximum WTP estimates translate into revenues? Maximum WTP should not be regarded as equal to an average payment that can be charged to the general public to pay for restoration. Often, when local funds are used to pay for stream restoration, this funding comes from government taxes or fees that all residents must pay. Thus, any taxes or fees necessary to pay for stream restoration must be acceptable to the majority of all residents. In the case of Deckers Creek, these taxes or fees would need to be set at a level that would gain the majority approval of the two-thirds of households estimated to have no positive value from stream restoration. Thus, these taxes or fees undoubtedly would need to be much lower than the maximum WTP estimates.
- The maximum WTP estimates depend on the type of market presented to respondents in an SCM question in addition to the restoration outcome.

In this study, Deckers Creek restoration was presented to survey respondents as a market with utility bill increases. It is hard to separate these two elements of the SCM question. Market type becomes an issue when many respondents reject the SCM question because of the market conditions presented.

- Survey data come from a sample of the population. In this case, telephone contact to obtain a sample was found to be an effective means from both response rate and cost perspectives. In addition, because stream users were found to be similar to the general population in their estimated WTP when using SCM, it may not be necessary to survey both groups separately.

- When using averages from sample data, analysts should be conservative when aggregating up estimates of maximum WTP to the entire population. Aggregation is needed to determine the benefit from stream restoration to the entire population. Conservative estimates treat non-respondents to the survey as having a zero monetary value for stream restoration.

- The general public cares more about outcomes from stream restoration than ecological measures used to quantify stream health. They want restoration to make the stream look clean and be safe for fishing or swimming. Thus, an SCM question must include policy-relevant descriptions of restoration outcomes, not just ecological measures (Adamowicz et al. 1997).

APPENDIX 3A: EXAMPLE OF THE PAPER SURVEY USED ON DECKERS CREEK

INTRODUCTION

This survey is being conducted by Alan Collins and Randy Rosenberger of the Agricultural and Resource Economics program at West Virginia University. The objective of this survey is to determine your attitudes and opinions about restoring the water quality of Deckers Creek in Monongalia and Preston Counties. Please answer the following questions to the best of your ability. You do not have to answer every question in this survey, and your participation is completely voluntary. Your opinions about stream water quality and Deckers Creek are greatly valued. All information gathered in this survey will be kept confidential. The only data released to the public will be in a form where no individual responses are identified.

There are a total of 30 questions in this survey, and it should take about 15 minutes to complete.

First, we would like to find out some general information about your outdoor recreation activities and your concerns about water quality of streams in West Virginia.

1. **Since May of 2001,** which of the following outdoor activities have you participated in? Please circle "yes" or "no" for each activity.

A. Fished in a lake, river, or creek	yes	no
B. Swam in a lake, river, or creek	yes	no
C. Explored or waded along a river or creek	yes	no
D. Kayaked or canoed	yes	no
E. Hunted	yes	no
F. Hiked or viewed wildlife	yes	no
G. Used a rail trail	yes	no
H. Had a picnic in the outdoors	yes	no

2. How important are each of the following aspects of streams to you? (Please select on a scale from 1 to 5 with 1 = very important; 3 = neutral; 5 = not important at all.)

- Streams that are safe to swim or wade in   __ 1   __ 2   __ 3   __ 4   __ 5
- Streams that are free of trash   __ 1   __ 2   __ 3   __ 4   __ 5
- Streams that are accessible for recreation   __ 1   __ 2   __ 3   __ 4   __ 5
- Streams that support fish and aquatic life   __ 1   __ 2   __ 3   __ 4   __ 5

3. In your opinion, how widespread are the following pollution sources of streams and rivers in West Virginia? (Please select on a scale from 1 to 5 with 1 = very widespread; 3 = somewhat widespread; 5 = not widespread at all.)

- Sewage   __ 1   __ 2   __ 3   __ 4   __ 5
- Acid and minerals from coal mining   __ 1   __ 2   __ 3   __ 4   __ 5
- Trash   __ 1   __ 2   __ 3   __ 4   __ 5

4. There are many reasons for restoring the water quality of polluted streams. For each of the reasons listed below, which category best describes how important each is from your point of view? (Please check one level for each.)

Reasons	Very important	Somewhat important	Not important
A. Providing habitat for fish and aquatic life.	____	____	____
B. Expanding recreation opportunities (fishing, swimming, picnicking, sightseeing, exploring, etc.) provided by a clean stream.	____	____	____
C. The pride and enjoyment that a clean stream provides to communities along it.	____	____	____
D. Knowing that future generations will be able to enjoy a clean stream.	____	____	____

Now, we would like to ask you some questions about Deckers Creek in Monongalia and Preston Counties.

5. What portions of Deckers Creek are you familiar with in terms of having used or seen these portions of the creek before? (Please check all that apply.)

_____ The lower portion of Deckers Creek as it flows from the Dellslow area to the Monongahela River

_____ The middle portion of Deckers Creek as it flows from Masontown to the Dellslow area

_____ The upper portion of Deckers Creek above Masontown

_____ I am not familiar with any portion of Deckers Creek

For more location information, see map insert of Deckers Creek Watershed.

6. **Since May of 2001,** how often have you used Deckers Creek or the rail trail alongside the creek?

___ None	___ 16–20 visits
___ 1–2 visits	___ 21–30 visits
___ 3–5 visits	___ 31–40 visits
___ 6–10 visits	___ 41–50 visits
___ 11–15 visits	___ more than 50 visits

7. **Since May of 2001,** have you fished, swam, waded, explored, kayaked, or canoed in Deckers Creek? (Please check one)

_____ yes _____ no

8. Based upon what you know about Deckers Creek, do you think there are environmental problems associated with it? (Please check one)

_____ yes

_____ no (skip to next page)

_____ don't know (skip to next page)

If you answered yes to Question 8, based upon what you know about Deckers Creek, what do you think are the main environmental problems associated with this creek? (Please check all that apply)

_____ Unnatural colors of the water and rocks along the creek

_____ Odor from the creek

_____ The lack of fish or aquatic life

_____ Trash in the creek and along the banks

_____ Unsafe to swim or wade in the creek

_____ Unsightly development along the creek

_____ High levels of acid and minerals in the water

_____ Other _____

Valuation Section

This next section will ask you to choose between the current condition of Deckers Creek and future possible conditions of a restored Deckers Creek. We will first define three traits of the creek—aquatic life, swimming safety, and scenic quality. In talking to people in the community, we have found that these are the most important traits of Deckers Creek. Each trait will have three levels (low, moderate, and high) describing the current condition of Deckers Creek and future possible conditions of a restored Deckers Creek.

We will use traits as defined below to describe current and future possible conditions of Deckers Creek.

Trait 1: Aquatic Life

This trait measures the ability to support aquatic life, including fish. We will use three levels to define this trait. Technical descriptions of the three aquatic life levels can be found in the provided insert.

Low   Current Condition   No Fish Habitat
             Explanation: This is the current condition of Deckers Creek with very limited areas of fishery
             habitat primarily due to acid and minerals (iron and aluminum) in the water.
Moderate Future Condition   Habitat to Support Stocked Fish
             Explanation: This is a future possible condition of Deckers Creek in which the water quality
             would improve enough to support stocking of trout in the middle portion of the creek
             (between Masontown and Dellslow).
High   Future Condition   Habitat for Reproducing Fish Populations
             Explanation: This is a future possible condition of Deckers Creek in which the water quality
             and stream habitat are improved such that sustained, reproducing fish populations could be
             established along the entire length of the creek. There would be trout populations in the
             middle portion and warm water fish (bass) in the lower portion of the creek (from Dellslow
             to the Monongahela River). This would include creation of enhanced fishery habitat for
             naturally producing populations in the lower portion of the creek.

9. How important is the aquatic life trait of Deckers Creek to you? (Please check one.)

_____ Very Important   _____ Somewhat Important   _____ Not Important

Trait 2: Swimming Safety

This trait measures the ability to safely swim or wade in Deckers Creek. We will use three levels to define this trait. Technical descriptions of the three swimming safety levels can be found in the provided insert.

Low   Current Condition   No Swimming
             Explanation: This is the current condition of Deckers Creek of unsafe water for swimming due to
             septic and sewage overflow. Staining, discoloration, and acidic water also create unpleasant
             swimming conditions.
Moderate Future Condition   Safe Swimming
             Explanation: This is a future possible condition of Deckers Creek in which the entire creek
             length meets the water quality standards for bacteria and is safe for swimming and wading.
             Municipal discharges (Morgantown and Masontown) of sewage are treated prior to release.
             The water is no longer acidic, but staining and discoloration in the creek still exist.
High   Future Condition   Safe, Enjoyable Swimming
             Explanation: This is a future possible condition of Deckers Creek in which the entire creek
             length exceeds the water quality standards for bacteria and is safe for swimming and wading.
             No untreated sewage from any source is discharged into the creek. The water is no longer
             acidic, and the staining and discoloration of the creek bed are cleaned up.

10. How important is the swimming safety trait of Deckers Creek to you? (Please check one.)

_____ Very Important   _____ Somewhat Important   _____ Not Important

Trait 3: Scenic Quality

This trait measures the natural beauty of the creek and surrounding banks. We will use three levels to define this trait:

Low   Current Condition   Some Litter

       Explanation: This is the current condition of Deckers Creek with periodic litter cleanups by volunteer groups.

Moderate   Future Condition   No Litter

       Explanation: This is a future possible condition of Deckers Creek where regular removal of all trash occurs from the stream and creek banks.

High   Future Condition   Creek Beautification and No Litter

       Explanation: This is a future possible condition of Deckers Creek where regular removal of all trash occurs from the stream and creek banks plus beautification of stream bank development is done along the lower part of the creek from Dellslow to the Monongahela River. This beautification would include trash receptacles along the rail trail and vegetative and flower planting plus erosion control along the banks of the creek where needed.

11. How important is the scenic quality trait of Deckers Creek to you? (Please check one.)

_____ Very Important   _____ Somewhat Important   _____ Not Important

EXPLANATION FOR CHOICE QUESTIONS 12–15

The next four questions will present you with options from which to choose. These choices will require you to make trade-offs between varying levels of the traits for Deckers Creek (including aquatic life, swimming safety, and scenic quality) that we just introduced and an associated cost. The cost, or price for cleaning Deckers Creek, will be charged as an increase in your monthly utility bill. The options given will include the current condition of Deckers Creek and two potential future conditions after restoration. **In each question, please choose the option (A or B) or current condition that best reflects how you feel about and what you would be willing to pay for restoration of Deckers Creek.**

Please treat these choices as if you were actually being offered the opportunity to restore Deckers Creek. We realize that any options you select are not available at the current time, but your choices are important in providing information to state environmental regulators and public officials about how the general public values stream restoration.

The levels of the traits in the potential future conditions are randomly assigned. We also include a price of restoration as an increase in your monthly utility bill (water, sewer, and/or electricity). The total revenue raised from increases in everyone's monthly utility bill would be used to finance the restoration of Deckers Creek.

Please consider the levels of the traits when selecting your most preferred option. Also, please make each choice independent of the ones preceding it.

12. **Choice 1**. Please consider the different trait levels and select the current condition or your most preferred option. Please check your choice below the table.

Trait	Current Conditions	Option A	Option B
Aquatic life	Low	High	High
Swimming safety	Low	Moderate	Moderate
Scenic quality	Low	High	High
Increase in monthly utility bill	$0	$2	$16
Please check one	_____	_____	_____

13. **Choice 2**. Please consider the different trait levels and select the current condition or your most preferred option. Please make this choice independent of the preceding choice question.

Trait	Current conditions	Option A	Option B
Aquatic life	Low	Moderate	High
Swimming safety	Low	Moderate	Moderate
Scenic quality	Low	High	High
Increase in monthly utility bill	$0	$4	$1
Please check one	_____	_____	_____

14. **Choice 3**. Please consider the different trait levels and select the current condition or your most preferred option. Please make this choice independent of the preceding choice question.

Trait	Current conditions	Option A	Option B
Aquatic life	Low	High	Moderate
Swimming safety	Low	High	High
Scenic quality	Low	High	High
Increase in monthly utility bill	$0	$2	$2
Please check one	_____	_____	_____

15. **Choice 4**. Please consider the different trait levels and select the current condition or your most preferred option. Please make this choice independent of the preceding choice question.

Trait	Current Conditions	Option A	Option B
Aquatic life	Low	High	High
Swimming safety	Low	Moderate	High
Scenic quality	Low	Moderate	Moderate
Increase in monthly utility bill	$0	$4	$2
Please check one	_____	_____	_____

16. Thinking about your responses to the previous four choice questions, please indicate how strongly you agree or disagree with each of the following (please check one level for each):

	1. Strongly agree	2. Agree	3. Neutral/ unsure	4. Disagree	5. Strongly disagree
A. I thought it was difficult to choose from among the options provided.	____ 1	____ 2	____ 3	____ 4	____ 5
B. I didn't have enough information to decide which option to choose.	____ 1	____ 2	____ 3	____ 4	____ 5
C. I don't think I should have to pay for restoration of Deckers Creek.	____ 1	____ 2	____ 3	____ 4	____ 5
D. I don't think Deckers Creek can be restored using only local funds.	____ 1	____ 2	____ 3	____ 4	____ 5
E. I am confident that I would have picked the same answers in Choices 1 through 4 if I was actually offered the opportunity to improve Deckers Creek.	____ 1	____ 2	____ 3	____ 4	____ 5

17. How do you think money should be collected in order to pay for restoring Deckers Creek? (Please check all that apply.)
    - _____ Utility bill increases
    - _____ Increase in local taxes
    - _____ Increase in state taxes
    - _____ Donations into specially designated trust funds
    - _____ Use existing government tax revenue and reduce spending elsewhere
    - _____ Don't know

18. Who do you think should be responsible for restoring Deckers Creek? (Please check all that apply.)
    - _____ Local communities
    - _____ People who would use Deckers Creek
    - _____ People who pollute Deckers Creek
    - _____ Friends of Deckers Creek (a watershed volunteer group)
    - _____ County government
    - _____ State government

_____ Federal government
_____ Other, please specify_____

We would like to finish this survey with some questions about you. These questions are for research purposes only. The information that you provide will remain confidential and will not be shared with any business or other institution. (Please check one response for each question.)

What is your gender?
_____ Male
_____ Female
What is your age?
_____ 18 to 25   _____ 46 to 55
_____ 26 to 35   _____ 56 to 65
_____ 36 to 45   _____ over 65
What is your race?
_____ African American   _____ Asian
_____ Caucasian (white)   _____ Hispanic
_____ Other
What is your current housing status?
_____ Homeowner
_____ Renter
_____ Other
Approximately how far away from Deckers Creek do you live?
_____ Alongside the creek
_____ Within sight of the creek
_____ Within 1 mile of the creek
_____ Between 1 and 5 miles from the creek
_____ Over 5 miles from the creek
How is your household sewage taken care of?
_____ City sewer system
_____ Household septic system
_____ Other _____
_____ Don't know
Do any children (17 years of age or younger) live in your household?
_____ Yes
_____ No
Including yourself, how many people live in your household? _____
What is your current employment status? (Please check one.)
_____ Employed full time
_____ Employed part time
_____ Student and part-time employment
_____ Full-time student
_____ Unemployed/looking for work
_____ Retired
_____ Home duties (homemaker)
_____ Self-employed
_____ Other (specify) _____

What was your total household income for 2000?

_____ Under $10,000	_____ $50,000–$60,000
_____ $10,000–$20,000	_____ $60,000–$70,000
_____ $20,000–$30,000	_____ $70,000–$80,000
_____ $30,000–$40,000	_____ $80,000–$90,000
_____ $40,000–$50,000	_____ $90,000–$100,000
_____ Over $100,000	

What is the highest level of education that you completed?

_____ Eighth grade or less	_____ Some college/technical school
_____ Some high school	_____ College degree
_____ High school diploma or GED	_____ Graduate school

Have you ever volunteered your time to participate in a stream cleanup project (either at Deckers Creek or at another stream or river)?

_____ yes   _____ no

Thank you for participating in this survey. Your time is appreciated. Please return this survey in the stamped, addressed envelope provided.

If you would like to be sent a written report of the results, please fill in your name and address below.

APPENDIX 3B: ATTRIBUTE DESCRIPTIONS FOR RESTORATION OF DECKERS CREEK

Aquatic Attribute: Ability to Support Aquatic Life, Including Fish

Levels:

(a) Low—Maintain status quo of very limited areas of fish habitat.
(b) Moderate—The water quality would be sufficient enough to support stocking of fish along the entire length of the creek (a put-and-take fishery). Warm-water species such as bass could be placed in the lower portion and cold-water species in the middle portion (trout).
(c) Complete—The water quality and stream habitat are improved such that sustained, reproducing fish populations are established along the entire length of the creek. This would include creation of enhanced fishery habitat for naturally producing populations in the lower part of Deckers Creek from Dellslow to the Monongahela River.

Scenic Attribute: Aesthetic Quality of the Creek and Surrounding Banks

Levels:

(a) Low—The status quo level of periodic litter cleanups by volunteer groups.
(b) Moderate—Regular removal of all trash from the stream and creek banks.
(c) Complete—Regular removal of all trash from the stream and creek banks plus beautification of stream bank development along the lower part of Deckers Creek from Dellslow to the Monongahela River. This

beautification would include trash receptacles along the rail trail and veg-
etative planting plus erosion control along the banks where needed.

Swimming Attribute: Ability to Safely Swim or Wade in the Water

Levels:

(a) Low—The status quo of unsafe for swimming due to septic and sewage
overflow discharges. Staining, discoloration, and acidic water also create
unpleasant swimming conditions.
(b) Moderate—The entire creek length meets the water quality standards
for bacteria and is safe for swimming and wading. Municipal discharges
(Morgantown and Masontown) of sewage are treated prior to release. No
more staining, discoloration, or acidic water exists.
(c) Complete—The entire creek length exceeds the water quality standards
for bacteria and is safe for swimming and wading. No untreated sewage
from any source is discharged into the creek. No more staining, discoloration, or acidic water exists.

Cost Attribute: Additional Monthly Cost per Household to Pay for Stream Restoration Included in the Utility Bill

Levels:

(a) Current: $0
(b) $1 per month
(c) $2 per month
(d) $4 per month
(e) $8 per month
(f) $16 per month

REFERENCES

Adamowicz, A., J. Swait, P. Boxall, J. Louviere, and M. Williams. 1997. Perceptions versus
objective measures of environmental quality in combined revealed and stated preference models of environmental valuation. *Journal of Environmental Economics and
Management* 32(1): 65–84.

Champ, P.A., K.J. Boyle, and T.C. Brown (Eds.). 2003. *A Primer on Nonmarket Valuation.*
Dordrecht, The Netherlands: Kluwer Academic.

Collins, A., R. Rosenberger, and J. Fletcher. 2005. The economic value of stream restoration.
Water Resources Research 41, W02017, doi:10.1029/ 2004WR003353.

Dillman, D.A. 2000. *Mail and Internet Surveys: The Tailored Design Method.* New York: Wiley.

Farber, S., and B. Griner. 2000. Valuing watershed quality improvements using conjoint analysis. *Ecological Economics* 34:63–76.

Krueger, R.A. 1994. *Focus Groups: A Practical Guide for Applied Research.* 2nd ed. Thousand
Oaks, CA: Sage.

Louviere, J.J., D.A. Hensher, and J.D. Swait. 2000. *State Choice Methods: Analysis and Applications.* New York: Cambridge University Press.

Mitchell, R.C., and R.T. Carson. 1989. *Using Surveys to Value Public Goods: The Contingent Valuation Method.* Washington, DC: Resources for the Future.

National Research Council. 2005. *Valuing Ecosystem Services: Toward Better Environmental Decision-Making.* Washington, DC: National Academies Press.

Stewart, J., and J. Skousen. 2003. Water quality changes in a polluted stream over a twenty-five-year period. *Journal of Environmental Quality* 32(2): 654–661.

Street, K. 2005. A Comparison of survey method effectiveness: Deckers Creek Survey v. Cheat River Survey. Senior thesis, West Virginia University, Morgantown.

USEPA. 2002. Mid-Atlantic acidification. http://www.epa.gov/region03/acidification/. Accessed May 6, 2003.

4 Using Hedonic Modeling to Value AMD Remediation in the Cheat River Watershed

James M. Williamson and Hale W. Thurston

CONTENTS

INTRODUCTION

States with active and abandoned mines face large private and public costs to remediate damage to streams and rivers from acid mine drainage (AMD). Calculating the cost of damage to streams and rivers due to AMD is not straightforward and can encompass a wide spectrum of factors, and as we have seen in Chapters 1 through 3 there are different ways to value these factors. There is the dollar loss of recreational activity associated with the streams and rivers, such as sport fishing and river rafting. In addition, other nonmarket values must be considered, although they might be more difficult to quantify. The hedonic pricing method, a revealed preference approach, takes advantage of the fact that sometimes nonmarket values are embodied in the price of other goods or services and therefore can be derived. Based on work developed by Rosen (1974), the implicit value of an attribute, such as water quality, is revealed by the observed price in a market transaction.

Using the hedonic pricing method, we focus on willingness to pay (WTP) for the cleanup of AMD-impaired waterways in the Cheat River watershed of West Virginia. We derive values for AMD using 21 years of housing sales data, spanning 1985–2005, and use geographic information systems (GIS) to link housing market sales data with stream water quality. The results indicate being located near an

AMD-impaired stream has an implicit marginal cost of between $5,023 and $6,044. We also find that the farther the house is from the stream, the smaller the effect of the amenity on price, but once houses are farther than about ¼ mile from a stream, they are not affected in a statistically significant way.

In the next section, we review the use of hedonic price models in environmental amenity valuation studies. Next, we describe our study area, the Cheat River watershed of West Virginia, followed by a description of our price and housing characteristic data and environmental water quality data. We then lay out our hedonic price model and address common econometric issues associated with price data with a spatial aspect. Finally, we report the WTP estimates for the watershed and conclude with a discussion of their impact on the remediation decision.

BACKGROUND

Hedonic price models are often used for deriving implicit values for unquantifiable or immeasurable characteristics. For example, Palmquist et al. (1997) used a hedonic price model to show that the odor associated with large-scale hog operations has a negative effect on the value of surrounding house sales. Their research revealed that odor, although not thought of as a good for which a market exists, is implicitly priced through the housing market. Research on air pollution has a long history of employing hedonic methods to derive the value of living in a neighborhood with high- or low-quality air (Smith and Huang 1995). In both cases, the factor that necessitates the use of hedonic modeling is the lack of a functioning market for the environmental good or service in question.

Economists describe houses as a set of attributes or characteristics and assume individuals have preferences for those attributes. These attributes describe the house and location (e.g., number of bathrooms, number of bedrooms, distance to parks, distance to schools, etc.) and surrounding environment (e.g., air and water quality). If economists can control for the house and location with data in the model, then the only reason for differences in home prices would be the environmental quality. The economic theory behind the implicit price representing value is stated by Rich and Moffitt (1982):

"The premise underlying the use of land value changes as a proxy for non-market benefits of environmental improvements to homeowners is that land values are bid up to the point where the marginal value of the improved water quality is equal to the marginal benefit" (p. 1033).

Similarly, the lack of a market for impaired streams and rivers creates the conundrum of how to value water quality. Assuming the amenities of clean water are embodied in the value of housing, we can use hedonic analysis.

Poor et al. (2007) used a hedonic framework to measure the marginal implicit price of nonpoint source water pollution. They found a relative implicit price of between 1% and 10% of the value of the average house, depending on the type of pollution.

STUDY AREA

The Cheat River watershed drains about 4,000 square kilometers and is one of the larger tributaries of the Monongahela River (Hansen et al. 2004). West Virginia's Department of Environmental Protection (WV-DEP) 303(d) list, so called because of the Clean Water Act (CWA) section requiring it, is a listing of impaired stream segments in the state. The listing reveals that many stream segments and tributaries of the Cheat River watershed and 115 kilometers of the Cheat River main stem are impaired or do not meet their designated uses (WV-DEP 2004). Much of the impairment is due to the legacy of coal mining (Williams et al. 1999; USEPA 2002).

States and authorized tribes must establish water quality standards to comply with the Clean Water Act. Water quality standards are made up of designated uses, or water quality goals, water quality criteria that protect the designated uses, an antidegradation policy to prevent backsliding, and general provisions for implementation (e.g., low flows, variances, etc.) (USEPA 1994).

Because of its tourism and recreational potential, including whitewater rafting in the Cheat Canyon region and recreational fishing, many studies have examined the possibility of restoring the Cheat River and its tributaries (e.g., Collins et al. 2005; Pavlick et al. 2005; Hansen et al. 2004; Williams et al. 1999; Ziemkiewicz et al. 2003). We focus on three subregions of the Cheat River watershed: Albright, Blackwater River, and Cheat River.

Within this geographic region, AMD occurs in several major tributaries, all of which are on the 303(d) list. In the Albright subregion, there are four major tributaries: Pringle Run (19.5 impaired kilometers), Lick Run (6.7 impaired kilometers), Heather Run (7.3 impaired kilometers), and Morgan Run (11.7 impaired kilometers) (WV-DEP 2004). Beaver Creek (20 impaired kilometers), located in the Blackwater River subregion, also suffers from AMD. Beaver Creek flows into the Blackwater River, which in turn flows into the Cheat River. In addition to the major tributaries, major portions of the Cheat River as well as many other smaller streams suffer from AMD impairment. Figure 4.1 illustrates the extent of stream impairment in the

Legend
— Impaired Streams
— Cheat River and Headwaters

FIGURE 4.1   Cheat River watershed with impaired streams.

Cheat River watershed. The streams in bold indicate stream impairment, and the thin lines indicate unimpaired streams.

The market or geographic set of people who would benefit from restoration of all or some of the waterways include those who would use the streams for fishing, water recreation, as a housing amenity, or simply value the resource's existence (i.e., non-use value). To make aggregate calculations of the WTP for restoration in this region, we delineate the market as the two counties, Preston and Tucker, or a subregion of the watershed. There were an estimated 14,544 households in the two counties according to the 2000 U.S. Census data.

Table 4.1 presents the characteristics of the Cheat River watershed. Although we focus on the two counties (Preston and Tucker), the watershed technically encompasses tracts of eight counties in West Virginia (U.S. Environmental Protection Agency [USEPA] 2005). Therefore, census estimates of the watershed's population, density, income, and so on are based on an average for the two counties where each county has an equal weight. Approximately 99% of the residents in the watershed were white, and the average household size was approximately 2.5 people. The residents had a median age of 41 years, 6 years older than the median age of U.S.

TABLE 4.1
Cheat River Watershed

Characteristic	Mean or Percent
Sex (% male)	49%
Race (% white)	98.85%
Age	41
Income (1990 $)	$20,568
Number of households	14,544
Household size	2.5[b]
Population	36,867
Population density	32/mi^2
Percent owner occupied	83%
Median house value (1990 $)	$62,100[a]

Sources: U.S. Census of Housing, General Housing Characteristics;
United States Census Bureau 2000 Census.

[a] From the U.S. Census Bureau, Statistical Abstract 1995.

[b] Mean value.

residents, and slightly more than half were women. Using U.S. Census figures, we report a weighted density of 32 residents per square mile, and the median income was $20,568 (reported in 1990 dollars).

The Consumer Price Index (CPI) is a measure of the average change over time in the prices paid by urban consumers for a fixed market basket of goods and services. It is a standard measure of inflation. To adjust all the prices in our data set to be able to compare them to a base year (in our case 1990) we use the formula: (price in whatever year) times (CPI in 1990 divided by CPI in whatever year). So for 1984, for example, if a house sold for $120,000 we use the formula $120,000 \times (127.9/103.8) = \$147,861$ The house that sold for 120,000 in 1984 has a price tag of $147,861 in 1990 dollars. http://www.bls.gov/cpi/

DATA

The data come from three sources. Data on housing characteristics as well as the sale price were collected from Preston and Tucker counties. These data were difficult to obtain. Most states and counties in the country have Web sites for the auditor at the county or other municipal level; information on housing sales is publicly available. Sometimes, the data are easy to download, and sometimes you have to write to get permission or a code to be able to download it. In the two counties of interest, the data were publicly available, but only accessible on 3 × 5 cards housed by the clerk in each county. The data were not in electronic format and required manual collection and input. Fortunately, we were able to contract someone to copy many of these transactions. Environmental stream quality data came from the WV-DEP Division of Water and Waste Management (www.wvdep.org). The data include streams that were listed as failing to meet the total maximum daily load (TMDL) guidelines and the stressors (e.g., pollutants) responsible for the streams' impairment. The stream data were geocoded, allowing us to combine water quality and housing market data using a geographic information system (GIS). For our study, we focused on one particular stressor, pH, as an indicator of AMD impairment. Finally, socioeconomic data from the counties came from the U.S. Census Bureau (http://www.census.gov/).

The years covered in the study are 1984 though 2005. Sale price for each observation was adjusted to reflect 1990 dollars. The structural characteristics allowed us to make comparisons of statistically similar houses. As stated, we needed to control for all the attributes (bedrooms, bathrooms, size in square feet, lot size, amenities like air conditioning, etc.) that contribute to the house price so we could focus on the environmental attribute of interest. The variable definitions and a summary of the data sample are provided in Table 4.2.

The final sample of data included 1,608 property sales over the 21-year period. We only included sales that were considered valid market sales, removing sales that

TABLE 4.2
Summary Statistics

Variable	Definition	Mean (Standard Deviation)
Sales price	Average sales price	$39,248 ($58,490)
AMD_0.25_miles	0–1 indicator of house located within ¼ mile of AMD impaired stream	0.222 (0.416)
AMD_0.50_miles	0–1 indicator of house located within ½ mile of AMD impaired stream	0.325 (0.468)
County	0–1 indicator for county; Preston County = 1	0.680 (0.466)
Central air conditioning	0–1 indicator for central air conditioning in house	0.067 (0.25)
Bedrooms	Number of bedrooms in house	2.78 (0.854)
Stories	Story height of house	1.27 (0.425)
Basement	0–1 indicator for basement	0.419 (0.494)
Heating	0–1 indicator for central heating	0.836 (0.370)
Bathrooms	Number of bathrooms in house	1.31 (0.675)
Sale acres	Acreage of sale	15.7 (37.1)
House size (ft²)	Square footage of house	1,389 (570)
Distance	Average distance to stream (miles)	0.489 (0.305)

were not at "arm's length." Further, we eliminated sales that were for land only to limit our analysis to housing.

The housing sales data (price and characteristics) were also geocoded. One important variable needed for the hedonic pricing method is the distance to the stream for each property. This distance calculation links each house not only to an impaired stream but also to the water quality of that stream. We used a buffer layer (Figure 4.2) using the geocoded water quality data. The buffer captured all properties within 1, ½, and ¼ mile of the stream. All of the properties in the sample were within 1 mile of a stream, whether it was impaired or not, and the average distance was 0.489 miles. Nearly one-third of the properties were within a ½ mile of an AMD-impaired stream, and one-fifth of the properties were within ¼ mile.

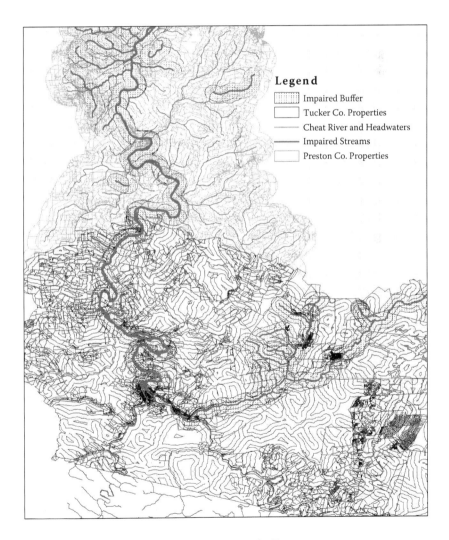

FIGURE 4.2   Cheat River watershed with property buffers.

MODEL

To estimate the value of water quality embodied in the sale price of a house, we use the following equation, called a semilogarithmic reduced form:

$$\ln P_{it} = \beta_0 + X_{it}\,\beta_1 + AMD_{it}\,\beta_2 + COUNTY\,\beta_3 + YRTREND\,\beta_4$$
$$+ MONTH\,\beta_5 + \varepsilon_{it} \tag{4.1}$$

One approach to deal with serial correlation would be to specify an autoregressive (AR) structure for the errors, but due to the nature of the sales data, it is infeasible. Not only is the panel unbalanced with unequally spaced sales across units, which does not itself preclude an AR approach, but a large share of the cross-section has only one sale in the 21-year sample period.

The dependent variable $\ln P_{it}$ is the natural log of the sales price of home i at time t. X_{it} is a vector of housing characteristics that include central air conditioning, number of bedrooms, story height, basement, central heating, number of bathrooms, acreage of the house's plot, and house size in square feet. The variable AMD is a 0–1 indicator for ¼-mile proximity to an AMD-impaired stream; therefore, β_2 is the marginal impact of water quality on sale price. $COUNTY$ is a 0–1 indicator for whether the sale took place in Preston or Tucker County. $YRTREND$ is a linear time trend that captures the trend in housing prices over the 21-year sample period. To account for the seasonal factor in sales, we include $MONTH$, which is a dummy for the month the sales took place. ε_{it} is the error term.

This section is pretty statistically heavy. If you're not familiar with some of the regression techniques (OLS, GLM) don't worry. The model building part of the hedonic method is important, but you needn't understand it *all* to get the gist of it.

Data for time series cross sections are notorious for containing regression errors that are not stable across time (Amacher and Hellerstein 1999). Nonnormally distributed error terms can be found in stock price data, fuel demand data, and wage and cost data. In our case, stream pollution also may be correlated with unobserved neighborhood characteristics in which case housing quality will be endogenous to water quality. Another potential source of bias comes from so-called emitter effects (see Leggett and Bockstael 2000, p. 124). Omitted variable bias produced by emitter effects is of particular importance in a hedonic analysis. For example, if the level of damage and the distance to a source are related, more variation is introduced into the exposure a homeowner experiences as the distance increases. The emitters in our research were abandoned mines and generally did not produce effects on other dimensions, such as air or noise pollution, that varied with the distance from the source. The emitting mines were located in rural mountainous areas, well hidden from view, and therefore were less likely to pose an "eyesore" to property owners or potential users of the contaminated streams. In fact, the location of many of the sources of AMD may have been unknown by the public.

Since we thought there might be these nonnormal errors, which would preclude us from relying much on our OLS (ordinary least-squares) estimate, we tested several models, starting with a basic OLS specification. We also tested a generalized linear model (GLM) with a correction for the unequal variance and individual specific random effects. Further, we allowed for serial correlation within housing units over time. Because of the unbalanced nature of our panel, which placed a constraint on the type of serial correlation we could specify, we used an exchangeable correlation structure and estimated a GLM.

RESULTS

The results of our hedonic regressions are presented in Table 4.3 (¼-mile indicator) and Table 4.4 (½-mile indicator). Column 1 of Table 4.3 presents the OLS model. The coefficient on the AMD variable indicated that houses located within a ¼ mile of an AMD impaired stream sold for 12.8% less, all else equal, than houses not located within a ¼ mile of an AMD-impaired stream. Calculating the difference in dollar terms, it was equal to $5,023, and the difference is statistically significant at the 5% level. Other factors that positively affected the sale price included number of bathrooms, acreage of the house plot, house size, and central heating. An additional bathroom added nearly 20%, or $7,850, to the sale price of a house. The linear trend variable indicates the sale price of houses in the two-county sample area rose by 2.6%, a value that is statistically significant at the 1% level. The coefficient on the basement variable indicates that the presence of a basement reduced the sale price, and the coefficient is statistically different from zero. The number of bedrooms and story height of the house also have negative signs, but we cannot say they are statistically different from zero.

The results from alternative estimators are similar to the OLS results. When we accounted for a randomly distributed individual specific constant term, the WTP coefficient on the AMD variable is −0.14 (standard error [SE] 0.069), a change of 15%. The maximum likelihood estimator (MLE) with error term corrections produced results nearly identical to the random effects model. In column 4 of Table 4.3, we relax the assumption that errors are independent across sales within housing units. When we cluster the error term by individual housing unit, the WTP coefficient is −0.143. Finally, accounting for autocorrelation in sale price with an exchangeable

TABLE 4.3
Regression Estimates for ¼-Mile AMD Impairment Buffer[a,b]

	OLS (1)	GLM with Random Effects (2)	MLE with Clustered Errors (3)	MLE with Autocorrelation (4)
AMD ¼ mile within house	−0.121*	−0.140*	−0.143**	−0.139*
	(0.056)	(0.069)	(0.062)	(0.067)
County	−0.192**	−0.144	−0.135	−0.144
	(0.056)	(0.080)	(0.075)	(0.078)
Central air conditioning	0.065	0.075	0.024	0.071
	(0.096)	(0.109)	(0.132)	(0.109)
Bedrooms	−0.022	−0.041	−0.031	−0.04
	(0.032)	(0.035)	(0.052)	(0.035)
Stories	−0.110	−0.103	−0.108	−0.102
	(0.064)	(0.068)	(0.081)	(0.067)
Basement	−0.208**	−0.259**	−0.220**	−0.257**
	(0.051)	(0.056)	(0.06)	(0.056)
Heating	0.283**	0.278**	0.230**	0.273**
	(0.076)	(0.081)	(0.09)	(0.08)
Bathrooms	0.179**	0.110*	0.156**	0.116**
	(0.047)	(0.050)	(0.056)	(0.05)
Sale acres	0.008**	0.008**	0.008**	0.008**
	(0.0007)	(0.0007)	(0.00015)	(0.0007)
House size (ft²)	0.0004**	0.0004**	0.0004**	0.0004**
	(0.00007)	(0.00006)	(0.00008)	(0.00006)
Year trend	0.026**	0.030**	0.027**	0.03**
	(0.004)	(0.004)	(0.005)	(0.004)
Constant	−40.60	−50.74	−44.67	−49.71
Model R^2	0.227	—	—	—
Partial R^2 of AMD variable	0.001			
F statistic (p value)	21.40			
	(p = .0000)	—	—	—
Breusch-Pagan (χ^2)	1.26			
	(p = .262)	—	—	—
Log likelihood	—	—	−2000.15	−1968.13
Observations	$N = 1,680$	$N = 1,503$	$N = 1,503$	$N = 1,503$

[a]Sample excludes sales of less than $1,000 and observations with missing acres.
[b]All prices have been adjusted to 1990 dollars.
**Significant at the 1% level; *Significant at the 5% level.

autocorrelation structure, the WTP is −0.139 (SE 0.067) (All commands were performed using Stata© software).

To test the sensitivity of our results to a change in distance to AMD impairment, we used a ½-mile indicator (see Table 4.4). The expanded buffer captured housing within ½ mile of an AMD-impaired stream. The coefficient that captured the effect

TABLE 4.4
Regression Estimates for ½-Mile AMD Impairment Buffer[a,b]

	OLS (1)	GLM with Random Effects (2)	MLE with Clustered Errors (3)	MLE with Autocorrelation (4)
AMD ½ mile within house	−0.074	−0.091	−0.112	−0.095
	(0.051)	(0.059)	(0.059)	(0.059)
County	−0.189**	−0.137	−0.128	−0.135
	(0.056)	(0.077)	(0.076)	(0.073)
Central air conditioning	0.061	0.068	0.025	0.061
	(0.096)	(0.109)	(0.131)	(0.109)
Bedrooms	−0.021	−0.039	−0.031	−0.037
	(0.032)	(0.035)	(0.052)	(0.035)
Stories	−0.112	−0.103	−0.108	−0.103
	(0.064)	(0.067)	(0.081)	(0.067)
Basement	−0.208**	−0.257**	−0.222**	−0.253**
	(0.051)	(0.055)	(0.059)	(0.055)
Heating	0.284**	0.272**	0.229**	0.263**
	(0.076)	(0.08)	(0.10)	(0.08)
Bathrooms	0.182**	0.119**	0.158**	0.128**
	(0.047)	(0.05)	(0.056)	(0.05)
Sale acres	0.008**	0.008**	0.008**	0.008**
	(0.0007)	(0.0007)	(0.0025)	(0.0007)
House size (ft^2)	0.0004**	0.0004**	0.0004**	0.0004**
	(0.00006)	(0.00006)	(0.00008)	(0.00006)
Year trend	0.026**	0.03**	0.027**	0.029**
	(0.004)	(0.004)	(0.005)	(0.004)
Constant	−41.74	−50.13	−45.03	−48.6
Model R^2	0.23	—	—	—
Partial R^2 of AMD variable	0.01			
F statistic (p value)	21.26			
	(p = .0000)	—	—	—
Breusch-Pagan (χ^2)	1.26			
	(p = .262)	—	—	—
Log likelihood	—	−1,969.20	−2,000.15	−1,968.13
Observations	N = 1,680	N = 1,503	N = 1,503	N = 1,503

[a] Sample excludes sales of less than $1,000 and observations with missing acres.

[b] All prices have been adjusted to 1990 dollars.

**Significant at the 1% level; *Significant at the 5% level.

of living within a ½ mile again indicated that there was an implicit cost embodied in the price of housing located near AMD-impaired streams for this region. The results for ½-mile distance also tell us that the effect of AMD on sale price was smaller, indicating a decreasing impact as distance from the pollution increased. In fact, the magnitudes of the WTP estimates for the ½-mile variable are as much as half the value of the ¼-mile WTP results. However, the effect was not statistically significant using any of the four estimators; thus, we cannot say the impact is different from zero.

We test the hedonic model for the presence of heteroskedasticity using the Breusch-Pagan test and we do not reject the null of constant variance (p-value = 0.262), therefore, the error term in our basic OLS model is unadjusted. The R^2 reported for the OLS suggests there still remains significant noise in the error, suggesting that our model has low explanatory power. Further, the partial R^2 of the AMD indicator is very small, which suggests the ability of the impairment variable to explain variation in sale price is a low approach, but a large share of the cross-section has only one sale in the 21-year sample period.

CONCLUSION

States with active and abandoned mines face large private and public costs to remediate damage to streams and rivers from AMD, but because of the lack of a market for these environmental goods or services, calculating the cost of damage to streams and rivers due to AMD is not straightforward. Our study contributes valuable information to the body of research on estimating the WTP for improved water quality. Using detailed housing and water quality data from a region of the Cheat River watershed that is heavily impaired by AMD, we used a hedonic price model to estimate an implicit cost of AMD damage. The innovation of our study is the use of a powerful tool, GIS, to connect a 21-year time series of housing sales with distance to impaired waterway.

Although most counties keep records of home sales and these home sales can help estimate changes in environmental attributes, the hedonic method requires difficult modeling techniques. These techniques may not be accessible to those without a statistical background; in fact, the names of the models and statistical tests (e.g., GLM and Breusch-Pagan test) can be confusing. If watershed groups feel that this revealed preference might be appropriate for their particular situation, economists and statisticians should be able to help collect and organize the data while also providing the expertise needed for a proper modeling effort.

The estimates from our hedonic model showed houses near AMD-impaired streams faced an implicit cost, although the farther the house was from the stream, the smaller the effect was. The results also showed that houses farther than ¼ mile from a stream were not affected in a statistically significant way. For watershed advocacy groups or others planning measures to restore streams affected by AMD, it is essential that they be aware of the many nonmarket benefits of their proposed restorations. Too often, the costs associated with restoration are highlighted, while the benefits are hidden. The hedonic method is particularly attractive because, unlike some other methods, it uses revealed preference data and estimates values that policy makers are likely to fully understand.

REFERENCES

Amacher, G., and D. Hellerstein. 1999. The error structure of time series cross section hedonic models with sporadic event timing and serial correlation. *Journal of Applied Econometrics* 14(3): 233–252.

Collins, A.R., R. Rosenberger, and J. Fletcher. 2005. Economic valuation of stream restoration. *Water Resources Research* 41:W02017–W0201.

Hansen, E., M. Christ, J. Fletcher, J. Petty, P. Ziemkiewicz, and R. Herd. 2004. The potential for water quality trading to help implement the Cheat Watershed acid mine drainage total maximum daily load in West Virginia. *Downstream Strategies* April.

Leggett, C.G., and N.R. Bockstael. 2000. Evidence of the effects of water quality on residential land prices. *Journal of Environmental Economics and Management* 39(2): 121–144.

Palmquist, R.B., F.M. Roka, and T. Vukina. 1997. Hog operations, environmental effects, and residential property values. *Land Economics* 73(1): 114–124.

Pavlick, M., E. Hansen, and M. Christ. 2005. Watershed based plan for the North Fork Blackwater River. *Downstream Strategies* August.

Poor, P.J., K.L. Pessagno, and R.W. Paul. 2007. Exploring the hedonic value of ambient water quality: A local watershed-based study. *Ecological Economics* 60(4): 797–806.

Rich, P.R., and L.J. Moffitt. 1982. Benefits of pollution control on Massachusetts' Housatonic River: A hedonic pricing approach. *Water Resources Bulletin* 18(6): 1033–1037.

Rosen, S. 1974. Hedonic prices and implicit markets: Product differentiation in pure competition. *Journal of Political Economy* 82(1): 34–55.

Smith, V.K. and J. Huang. 1995. Can markets value air quality? A meta-analysis of hedonic property value models. *Journal of Political Economy* 103(11): 209–227.

U.S. Environmental Protection Agency. 1994. *Water Quality Standards Handbook.* 2nd ed. EPA/823/B-94/005a. Washington, DC: U.S. Environmental Protection Agency, Office of Water. http://www.epa.gov/waterscience/library/wqstandards/handbook.pdf (accessed May 2005).

U.S. Environmental Protection Agency. 2002. Mid-Atlantic acidification report, Washington, D.C. http://www.epa.gov/region3/acidification/. (accessed January 2007).

U.S. Environmental Protection Agency. 2005. Watershed profile. http://cfpub.epa.gov/surf/huc.cfm?huc_code=05020004 (accessed May 2005).

Williams, D.R., M.E. Clark, and J.B. Brown. 1999. *Stream and Water Quality in Coal Mined Areas of the Lower Cheat River Basin, West Virginia and Pennsylvania During Low-Flow Conditions, July 1997.* U.S. Geological Survey Water-Resources Investigations Report 98-4258 (U.S. Department of the Interior: Washington, DC).

Ziemkiewicz, P., J. Skousen, and J. Simmons. 2003. Long-term performance of passive acid mine drainage treatment systems. *Mine Water and the Environment* 22: 118–129.

5 Using Benefit Transfer to Value Acid Mine Drainage Remediation in West Virginia

James M. Williamson, Hale W. Thurston, and Matthew T. Heberling

CONTENTS

INTRODUCTION

In this chapter, we demonstrate the use of benefit transfer (BT) techniques to estimate the benefits of restoring an impaired region of the Cheat River watershed in West Virginia. We identify a subregion of the watershed with substantial acid mine drainage (AMD) damage, as indicated by the total maximum daily load (TMDL) program's 303(d) list, and estimate the welfare benefits of restoration (measured in dollars, or "monetized") using the BT method.

BACKGROUND

The restoration of certain stream reaches is likely to be marginally more beneficial than others due to factors such as location, access, or recreational fishing quality. The costs are likely to be different as well.

Water quality standards are made up of designated uses or water quality goals, water quality criteria to support the designated uses, and antidegradation policy to prevent backsliding. Waters that do not meet their designated uses are said to be impaired and are placed on the U.S. Environmental Protection Agency's (USEPA's) 303(d) list.

The U.S. Environmental Protection Agency (USEPA) is increasingly interested in determining the monetized benefits of their restoration activities. In many cases, an economic analysis, usually in the form of a benefit-cost analysis, is required for proposed regulations. In 1993, President Clinton signed Executive Order 12866 on Regulatory Planning and Review (58 *Federal Register* 51735, October 4, 1993) that required agencies to perform a benefit-cost analysis if the regulatory actions met several criteria, including having an annual impact on the economy of $100 million. Even when it is not required, an economic analysis can provide a useful framework for integrating the expected public health, ecological, and other types of impacts of policies or restoration alternatives into a single overall measure and presenting those impacts in terms easily understood by decision makers and the public.

In the face of budget and time constraints, BT is a cost-effective valuation method that utilizes previous resource value estimates to make judgments about the value of resources at a policy site.

The results of our research provide policy makers with the necessary estimates to begin the prioritization of restoration projects. The next section describes the area of study, the Cheat River watershed in northeastern West Virginia, which is one of the most AMD-affected areas of the Mid-Atlantic Highland region.

THE STUDY REGION

The Cheat River watershed drains about 4000 square kilometers and is one of the larger tributaries of the Monongahela River. West Virginia's Department of Environmental Protection (WV-DEP) 303(d) list, so called because of the Clean Water Act (CWA) section requiring it, is a listing of impaired stream segments in the state. The listing reveals that the Cheat River watershed has a number of stream segments and tributaries that are impaired (WV-DEP 2004). Much of the impairment is due to the legacy of coal mining (Williams et al. 1999; USEPA 2002). Based on the WV-DEP 303(d) list, approximately 115 kilometers of the Cheat River main stem do not meet water quality goals.

Because of its tourism and recreational potential, including whitewater rafting in the Cheat Canyon region and recreational fishing, many studies have examined the possibility of restoring the Cheat River and its tributaries (e.g., Collins et al. 2005; Pavlick et al. 2005; Hansen et al. 2004; Williams et al. 1999; Ziemkiewicz et al. 2003). We focus on three subregions of the Cheat River watershed: Albright,

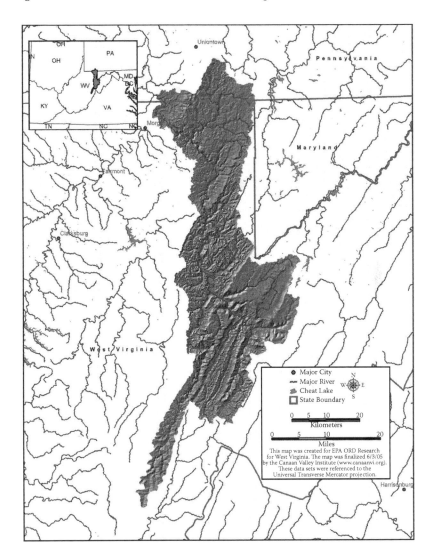

FIGURE 5.1   Cheat River watershed region map.

Blackwater River, and Cheat River. Figure 5.1 presents a map of the region. Within this geographic region, AMD occurs in several major tributaries, all of which are on the 303(d) list. In the Albright subregion, there are four major tributaries: Pringle Run (19.5 impaired kilometers), Lick Run (6.7 impaired kilometers), Heather Run (7.3 impaired kilometers), and Morgan Run (11.7 impaired kilometers) (WV-DEP 2004). Beaver Creek (20 impaired kilometers), located in the Blackwater River subregion, also suffers from AMD. Beaver Creek flows into the Blackwater River, which in turn flows into the Cheat River.

The market or geographic set of people who would benefit from restoration of all or some of the waterways can include those who would use the streams for fishing, water recreation, as a housing amenity, or for the resource's existence. To make aggregate calculations of the willingness to pay (WTP) for restoration of the region, we delineated the market as the two counties, Preston and Tucker, in West Virginia encompassing the subregions of the Cheat River watershed. There were an estimated 14,544 households in the two counties according to the 2000 U.S. Census data.

BENEFIT TRANSFER

Smith et al. (2000) noted that BT involves four general steps: (1) translate the policy change into one or more resulting quantity changes for uses that are linked to an environmental resource; (2) estimate the number of typical users before and after the policy change; (3) transfer a per "unit" consumer surplus measure, with the unit measure comparable to the index used in step 1; and (4) combine estimates from the first three steps for each year considered in the analysis and compute the discounted aggregate benefit measures. Scale issues will come into play primarily in the first two steps. In the first step, ecological scale effects must be taken into consideration, and in step 2, the researcher needs to determine market area or the number of users, which may change as the scale of the project goes from local to regional to national.

A recent development in the BT literature, known as *preference calibration,* offers a way of linking the fundamental preference structure with welfare measurements. The advantage of the calibrated preference over the unit benefit approach is that it allows for consistent values to be estimated based on underlying economic theory and available benefit estimates. This chapter presents the simplest BT analysis. A more complex analysis using preference calibration can be found in Williamson et al. (2007).

By linking preference functions with actual WTP estimates, researchers can "back out" the parameters of the function. These derived parameters, which Smith et al. (2002) referred to as "calibrated," can then be used in a new WTP function for our policy site.

METHODS

DIRECT TRANSFER

In our study, we used estimates of WTP from four major studies: two studies in which AMD was the pollutant or stressor (Farber and Griner 2000; Collins et al. 2005) and two studies of general water pollution (Smith and Desvousges 1986; Carson and Mitchell 1993). The transfer assumes that the policy and study sites have been judged as comparable. After deriving an average estimate of the household WTP for partial and total restoration, we aggregated the estimate using U.S. Census household figures for the watershed.

BT does not provide exact measures, and there is a good deal of debate in the literature surrounding the best methods for conducting the transfer. It has been suggested that transferring the functional form or estimating equation from a study site to the policy site is preferable to the simple transfer of raw benefit estimates. Loomis (1992) argued that a functional transfer has the benefit over a direct transfer because estimated coefficients are able to capture the specific characteristics of the policy site better than direct transfer. Many times, however, transferring estimated coefficients from a study site to a policy site is not feasible. We acknowledge that there are limitations to both methods of BT, and we base the quality of estimates on the soundness of the estimates from the study sites.

CHARACTERISTICS OF THE TRANSFER STUDIES

The number of benefit point estimates available to transfer was constrained by the number of studies that have been conducted. Having multiple studies to draw from is preferable to a single study because it enhances the researcher's ability to conduct a more meaningful transfer. Multiple studies allowed us to establish estimated bounds of the WTP, and the studies' estimated error structures made it possible to estimate confidence intervals for the transfer (Desvousges et al. 1998). We have identified two important studies that estimated the marginal value of restoration of AMD-impaired waterways and two studies in which general water pollution was the stressor. *General* refers to a nonspecific source or type of pollution. In this section, we provide a comprehensive analysis of the suitability of the studies for a BT to the Cheat watershed. We start by comparing the empirical methods used in the studies and then describe the watershed attributes.

Table 5.1 summarizes the critical aspects, including the WTP estimates, of each study. We drew WTP estimates from studies in which the objective was to restore a polluted or damaged water body. The first two AMD studies provided WTP estimates of restoration for a particular type of impairment in the Mid-Atlantic Highland region, the region in which the Cheat River watershed is also located. A third study conducted in the region surveyed the population's WTP for restoration of a large river basin for general water pollution. The fourth study we selected encompassed the entire United States and provided national estimates of WTP for general water quality. We used the national study as a measure to gauge the difference in WTP estimates where the pollution was local and immediate to the survey participants from those where the pollution was presented in a national context.

First, we note that all of the studies chosen for the transfer used a similar elicitation method. Carson and Mitchell (1993) used a contingent valuation approach to survey a sample drawn from the U.S. population. The two studies of AMD restoration presented residents with the issue of a watershed with environmentally damaged waterways, shared a common underlying survey method to sample users and nonusers in the watersheds, and used the stated choice method to elicit restoration values.

Further, Farber and Griner (2000) and Collins et al. (2005) reported the welfare implications for similar restoration outcomes. Farber and Griner described a

TABLE 5.1
Comparison of Willingness-to-Pay Studies

	Study			
	(1)	(2)	(3)	(4)
Feature	Smith and Desvousges (1986)	Carson and Mitchell (1993)	Farber and Griner (2000)	Collins et al. (2005)
Type of water impairment	General water pollution	General water pollution	Acid mine drainage	Acid mine drainage
Geographic scope	Regional river basin	National waterways	Regional watershed	Regional watershed
Number of households	2.5 million households	92 million households	242,416 households	35,719 households
Restorative definitions	Boatable, fishable, swimmable	Boatable, fishable, swimmable	Moderately polluted, unpolluted	Aquatic quality, scenic quality, swimming safety
WTP elicitation method	Contingent valuation	Contingent valuation	Stated choice method	Stated choice method
Sample size (*N*)	301	564	367	257
Annual household WTP for restoration (2004 dollars)				
Severe → moderate	$58.54 ($22.78)	$84.08[a,b] ($8.99)	$43.38 (7.45)	$75.88[a] ($21.60)
Moderate → unpolluted	$46.46 ($17.73)	$91.48[c] ($8.50)	$27.45 ($3.31)	$45.43[c] ($13.92)
Severe → unpolluted[d]	$105.00 ($20.26)	$175 ($8.75)	$77.00 ($5.36)	$121.31 ($17.76)

Note: Standard errors in parentheses.

[a] Average WTP for restoration from boatable to fishable.

[b] We have scaled the national estimates down to regional level by the factor of 0.7 used by Smith et al. (2002).

[c] Restoring water from fishable to swimmable.

[d] Full restoration values are estimated by a unique coefficient in Smith and Desvousges (1986) and Farber and Griner (2000); however, Collins et al. (2005) noted in their results (p. 71) that the sum of the marginal valuations for severe to moderate and moderate to unpolluted equals the valuation of severe to unpolluted. Carson and Mitchell (1993) explicitly used the sum of the marginal benefits to produce a total WTP figure (p. 2449).

scenario of restoration that moved the stream from moderately polluted to unpolluted, and Collins et al. stated the cleanup in terms of moderate or full restoration. Although Farber and Griner asked survey participants only to trade off different levels of restoration, the survey of Collins et al. allowed participants to make separate trade-offs for the levels of restoration for aquatic life, scenic quality, swimming safety, and cost. Farber and Griner respondents were choosing only restoration levels based on aquatic quality, but in our AMD study, if restoration to the level of "unpolluted" made reproduction among aquatic life possible, then it also made swimming and wading possible. Also, Farber and Griner respondents were not asked about their preferences for scenic quality, but we were less willing to assume respondent valuations embody scenic quality. Therefore, to account for this difference in the surveys, only the marginal values for the two attributes of swimming quality and aquatic life were used from Farber and Griner.

Table 5.2 compares U.S. Census figures from three regional watersheds and the national waterway with the characteristics of the survey respondents from each study. A side-by-side comparison of U.S. Census figures for the watersheds and survey respondent demographics allowed us to judge how representative the WTP estimates from the studies were of their respective watershed regions. Table 5.2 also presents the characteristics of the Cheat River watershed, our policy (or transfer) site.

The Cheat River is located in the same region as Deckers Creek, the lower Allegheny watershed, and the Monongahela River basin, so it should not be a surprise that they have very similar demographic characteristics. In our study, we restricted the definition of the boundaries for the watershed to two counties (Preston and Tucker), although the watershed technically encompasses tracts of eight counties in West Virginia (USEPA 2005). Census estimates of the watershed's population, density, income, and so on were based on an average for the two counties in which each county had an equal weight. The gender ratio was similar in all three watersheds. About 90% of the residents in the watersheds were white, and the average household size was approximately 2.5 people. The residents of the Cheat River watershed and lower Allegheny had a median age more than 7 years older than the median age of U.S. residents, whereas the residents of Deckers Creek, with a median age of 35, resembled the U.S. population. Residents of the Monongahela River basin were much older, on average, than the U.S. population and were 5 to 11 years older on average than individuals in the rest of the watersheds in our study.

We should also mention some other differences among the different watersheds. First, the population densities of the lower Allegheny and the Monongahela were much higher than either the Cheat or Deckers. Using U.S. Census figures and the watershed boundary definition of the Pennsylvania Environmental Council, we estimated a weighted density of 788 residents per square mile in the lower Allegheny, and Smith and Desvousges (1986) reported a density of 518 residents per square mile for the Monongahela. The Cheat River watershed, on the other hand, had a density of 32 residents per square mile, and Deckers Creek, while more densely populated than the Cheat River watershed, was still well below the density of the Allegheny and Monongahela. The difference in density figures is driven partly by the inclusion of Allegheny County in both of the high population density watersheds (33% of the lower Allegheny and more than 23% of the Monongahela River basin). Allegheny County is home to Pittsburgh, Pennsylvania, and the county alone has nearly the population of the entire state of West Virginia. Overall, the population density of the United States was 74 residents per square mile, placing it between the Cheat River watershed and Deckers Creek.

TABLE 5.2
Comparison of Characteristics by Study Respondents and U.S. Census Reports

Characteristic	Transfer Site Cheat River Watershed	U.S. Census A Monongahela River Basin	B National Waterways	C Lower Allegheny Watershed	D Deckers Creek	Survey Respondents A' Smith and Desvousges (1986)[a]	B' Carson and Mitchell (1993)[a]	C' Farber and Griner (2000)	D' Collins et al. (2005)
Sex (% male)	49%	47%	49%	48%	50%	37%	49%	68%	44%
Race (% white)	98.85%	92%	80%	94.2%	95.5%	90%	80%	99%	92%
Age	41	46	33	40	35	48	33	51[b]	45
Income (1990 $)	$20,568	$23,784	$27,050	$27,970	$19,480	$23,250	$27,050	$28,500	$30,500[b]
Number of households	14,544	890,000[c]	92 million	242,416	23,244				
Household size	2.5[b]	2.8[c]	2.6[b]	2.4[b]	2.4[b]	n.r.	2.6[b]	2.9[b]	2.38[b]
Population	36,867	2.5 million	248 million	576,885	55,785				
Population density	32/mi²	518/mi²	73.6/mi²	788/mi²	135.95/mi²				
Percent owner occupied	83%	71.5%	64%	74%	61%				
Median house value (1990 $)	$62,100	$59,440	$79,100	$84,287	$79,300	n.r.	$79,100	n.r.	n.r.

n.r. (not reported by the authors).

[a] From the U.S. Census Bureau, Statistical Abstract 1995.

[b] Mean value.

[c] From U.S. Census of Housing, General Housing Characteristics, United States Census Bureau 2000 Census.

Population density is important because it helps us distinguish between rural and metropolitan areas. Unlike the lower Allegheny and the Monongahela River basin, the Cheat River watershed includes only towns and small cities. In terms of meaningful differences for our transfer, the density figure may have implications for how the residents value water quality. For example, because residents sort themselves into rural and urban populations, there are systematic differences between them. There may be observable differences, such as income, that influenced the WTP for restoration, but there also may be less-obvious differences, such as tastes and preferences for environmental amenities. These differences, both observed and unobserved, will in turn affect residents' willingness and ability to pay for a restoration project.

It should not be a surprise that median household income also differed significantly among the watersheds given the difference in population density and the inclusion of Pittsburgh in the lower Allegheny and Monongahela River basin. Residents of the lower Allegheny had a median annual household income that was almost ten thousand dollars higher than the residents of the Cheat River Watershed or Deckers Creek. The median value of an owner-occupied home was also much higher in the lower Allegheny. The median value of houses in the Cheat River watershed and Deckers Creek were $62,100 and $79,300, respectively, whereas the median house in the Lower Allegheny was worth $84,287. Only the Monongahela River basin had a lower median housing value.

Although the studies took due care to draw a random sample from the populations in the study sites, the characteristics describing the population living in the watersheds did not necessarily represent the characteristics of those who responded to the WTP surveys, as shown in columns A' to D' of Table 5.2. Survey respondents differed in many respects from U.S. Census figures for their respective watershed. The median income of survey respondents in both studies was higher than that of the watershed as a whole, and in the case of Collins et al. (2005), the income was almost $13,000 higher. Further, the median income of survey respondents was significantly higher than the income of the Cheat River watershed population and slightly higher than the median income in the United States. The implication for the estimates is an increasing WTP among higher-income watersheds, based on previous empirical evidence that WTP increases with income (Poe et al. 2001). Further, our own meta-analysis of the four studies showed a small but positive effect of income on WTP.

It is important to keep in mind that WTP reflects both willingness and ability to pay. Rural residents or residents with low incomes may have tastes or preferences for environmental amenities that are masked by their ability to pay. Because WTP estimates are used to prioritize restoration projects, areas with lower incomes could be ranked lower, holding other things constant. Although not explicitly accounted for in BT, the issue of environmental justice or fairness is one that policy officials and stakeholders should also take into account when valuing projects.

A large majority of respondents in the Farber and Griner (2000) study were men (only about 30% of those who responded were women), and respondents were 10 years older than the median resident in either watershed. The differences between the characteristics of the survey respondents and the watershed populations likely resulted from response bias, and it is necessary to consider how the responses of those who chose to respond should be used to infer benefits of stream restoration in the Cheat River watershed. For example, because respondents to the Farber and Griner survey were from higher-income households, it is likely that they had higher levels of education. Also, the high proportion of men responding could also have had an effect on the valuation of benefits of restoration if men and women had different tastes for environmental amenities.

As we have shown, the two AMD studies produced the most applicable estimates for our BT study based on type of pollution, degree of impairment, similarity of the watershed, and resident populations. Both AMD studies were conducted in the same region of the United States as our policy site. Both used similar survey methods, and the surveys sampled a population that had similar demographic characteristics. The two additional studies, which respectively valued general water quality on a national and local scale, provided a metric against which we could examine the regional AMD welfare estimates. The national and regional studies compared well to the two regional AMD studies in terms of survey methods and empirical design and thus provided more information on the accuracy of our BT studies. Hence, for a direct transfer of mean benefits from the study sites to the policy site, we chose one study of general water pollution and two AMD studies from the Mid-Atlantic Highland Region and a national study. Each of the studies used a survey format to elicit WTP values and thus produced benefit estimates in a comparable manner.

RESULTS

We report the results of a BT method used to value water restoration at the Cheat River watershed policy site. The results for both models are in Table 5.3. Columns 1 and 3 present the results of the direct transfer of mean WTP estimates from the selected study sites.

Overall, the estimated WTP for the direct transfer was closely grouped across the four studies. Column 1 contains the means of a direct transfer using only the AMD studies. Annual household WTP figures, the amount a household would be willing to pay for an incremental change in water quality, ranged from $36 to $99. The estimates behaved predictably, displaying a decreasing marginal utility in remediation as the water quality increased.

Estimates of the aggregate value of partial restoration of waterways in the Cheat River watershed, derived as the product of the mean WTP and total number of households, were close to $1 million per year. If the waterways were to be totally restored, that is, a severely polluted stream cleaned up to a level at which it was capable of sustaining aquatic life, supporting fishing, and allowing human contact with the water, the total annual household WTP was $1.4 million.

When we incorporated the studies by Smith and Desvousges (1986) and Carson and Mitchell (1993) as part of the BT estimate, the WTP estimates increased for

TABLE 5.3
Estimated Remediation Benefits

	Direct Transfer		
	1	2	3
	AMD Studies Only	AMD and General Water Quality Studies	Studies Used in Preference Calibration Model
Water quality change			
Severe → moderate	$59	$65	$54
Moderate → unpolluted	$36	$53	$77
Severe → unpolluted	$99	$120	$131
Annual aggregate WTP for restoration[a,b,c,d]			
Severe → moderate	$0.9	$1.0	$0.8
Moderate → unpolluted	$0.5	$0.8	$1.1
Severe → unpolluted	$1.4	$1.8	$1.9

Note: All figures have been adjusted to 2004 dollars.
[a] Based on 14,544 households in the watershed.
[b] Partial restoration is defined as restoration of severely to moderately polluted or of moderately to unpolluted.
[c] Full restoration is defined as restoration of a severely polluted stream to unpolluted.
[d] Dollar figures are in millions.

partial and total restoration. The estimates in column 2 suggest that households were willing to pay relatively more to restore severely damaged streams. The WTP for restoring a stream from severely polluted to moderately polluted was $65, and they would be willing to pay $53 to completely restore a moderately polluted stream. The benefit estimate of fully restoring a severely polluted stream was $120. The aggregate household WTP for fully restoring a severely polluted stream (based on a direct transfer using all four studies) was $1.8 million.

Although due care was taken in a direct transfer to closely match study and policy sites, ultimately a mean transfer cannot explicitly incorporate the economic constraints like the preference calibration approach. Also, differences may arise because of the way water quality changes affect WTP estimates in a preference calibration. As noted by Smith et al. (2002), because the preference calibration approach is consistent with an individual's preference structure, calibrated WTP values always increase with incremental improvements in water quality.

CONCLUSION

This chapter presents a direct BT for a policy site in the Cheat River watershed of northeastern West Virginia. We drew from two previous studies that measured the WTP of residents in a watershed with AMD-impaired waterways. In addition to examining a common stressor, the studies shared the same region as our policy site watershed, an area referred to as the Mid-Atlantic Highlands. We also drew on two studies in which the stressor was not AMD, but instead freshwater pollution. These studies provided estimates of the WTP for restoration of an impaired body of water without a specific stressor being associated with the degradation. Again, we chose a study that was conducted in the Mid-Atlantic Highland region because it closely matched our policy site in terms of population characteristics and geography. A second general freshwater pollution study conducted throughout the United States that employed a representative sample of the entire population was also used.

Based on the estimates from the four study sites, remediation work in the Cheat River watershed had an annual household WTP of between $36 and $65 for partial restoration, depending on the studies included and the level of restoration. The household WTP for full restoration, using all of the studies, was $120. If we restricted our transfer to only the AMD studies, the expected annual household WTP was $99. Given the annual household WTP, we estimated a total benefit figure of between $1.4 and $1.8 million for total restoration in the Cheat watershed. Estimates from the preference calibration technique produced larger estimates of the WTP. Based on the results of the contingent valuation method literature, we generally expect the benefit of restoring a severely polluted stream to be greater than the benefit of fully restoring a moderately polluted stream.

West Virginia has thousands of kilometers of AMD-degraded streams that are in need of restoration. Given limited resources, state agencies, policy makers, and stakeholder groups need access to accurate benefit estimates to efficiently target restoration projects. By its nature, BT has attracted a fair amount of skepticism. Validity tests performed in recent work have shown BT estimates are susceptible to large variation (Delavan and Epp 2001; Vandenberg et al. 2001). Indeed, transferring estimated WTP values from a study site to a policy site requires a good deal of judgment on the part of researchers conducting the transfer and always opens the process up to errors or biases (Kirchhoff et al. 1997). While accepting the limitations of direct transfer, we also recognize the utility of employing the considerable work that has been conducted in the region with regard to valuation of ecological commodities. The value of BT is that it provides us with a baseline for welfare estimates of the restoration values in the Cheat watershed. By combining the estimated welfare value of restoration (measured in dollars) with the estimated cost of restoration for a particular stream segment, we were able to justify restoration projects. Comparing restoration value-to-cost ratios of different stream segments, we can help prioritize restoration projects.

REFERENCES

Carson, R.T., and R.C. Mitchell. 1993. The value of clean water: The public's willingness to pay for boatable, fishable, and swimmable quality water. *Water Resources Research* 29(7): 2445–2454.

Collins, A.R., R. Rosenberger, and J. Fletcher. 2005. Economic valuation of stream restoration. *Water Resources Research* 41: W02017-W0201.

Delavan, W., and D. Epp. 2001. Benefits transfer: The case of nitrate contamination in Pennsylvania, Georgia and Maine, in *The Economic Value of Water Quality*, Eds. J.C. Bergstrom, K.J. Boyle, and G.L. Poe, Northampton, MA: Elgar, pp. 121–136.

Devousges, W.H., F.R. Johnson, and H.S. Banzhaf. 1998. *Environmental Policy Analysis with Limited Information: Principles and Applications of the Transfer Method.* Cheltenham, U.K.: Elgar.

Executive Order no. 12866, Regulatory Planning and Review, *Federal Register* 58 (October 4, 1993): 51735.

Farber, S., and B. Griner. 2000. Valuing watershed quality improvements using conjoint analysis. *Ecological Economics* 34: 63–76.

Hansen, E., M. Christ, J. Fletcher, J. Petty, P. Ziemkiewicz, and R. Herd. 2004. The potential for water quality trading to help implement the Cheat Watershed acid mine drainage total maximum daily load in West Virginia. *Downstream Strategies* April.

Kirchhoff, S., B.G. Colby, and J.T. LaFrance. 1997. Evaluating the performance of benefit transfer: An empirical inquiry. *Journal of Environmental Economics and Management* 33: 75–93.

Loomis, J.B. 1992. The evolution of a more rigorous approach to benefit transfer: Benefit function transfer. *Water Resources Research.* 28(3): 701–705.

Pavlick, M., E. Hansen, and M. Christ. 2005. Watershed based plan for the Lower Cheat River Watershed: From river mile 43 at Rowlesburg, WV, to the West Virginia/Pennsylvania border, including all tributaries. *Downstream Strategies* February.

Poe, G.L., K.J. Boyle, and J.C. Bergstrom. 2001. A preliminary meta analysis of contingent values for ground water quality revisited, in *The Economic Value of Water Quality*, Eds. J.C. Bergstrom, K.J. Boyle, and G.L. Poe, Northampton, MA: Elgar, pp.137–162.

Smith, V.K., and W.H. Desvousges. 1986. *Measuring Water Quality Benefits.* Boston: Kluwer-Nijhoff.

Smith, V.K., G.L. Van Houtven, and S.K. Pattanyak. 2002. Benefit transfer via preference calibration: "Prudential Algebra" for policy. *Land Economics* 78(1): 132–152.

Smith, V.K., G.L. Van Houtven, S. Pattanyak, and T.H. Bingham. 2000. *Improving the Practice of Benefit Transfer: A Preference Calibrated Approach.* Interim Final Report to the U.S. Environmental Protection Agency, U.S. Environmental Protection Agency: Washington, DC.

U.S. Environmental Protection Agency, Region 3. 2001. *Metals and pH TMDLs for the Cheat River Watershed, West Virginia.* Prepared by Tetra Tec, Inc., Contract 68-C-99–24, U.S. Environmental Protection Agency: Washington, DC.

U.S. Environmental Protection Agency. 2002. Mid-Atlantic acidification report, Washington, DC. http://www.epa.gov/region 3/acidification/. (Accessed January 2007).

U.S. Environmental Protection Agency (USEPA). 2005. Watershed profile. http://cfpub.epa.gov/surf/huc.cfm?huc_code=05020004 (accessed May 2005).

Vandenberg, T.P., G.L. Poe, and J.R. Powell. 2001. Assessing the accuracy of benefits transfer: Evidence from a multi-site contingent valuation study of ground water quality, in *The Economic Value of Water Quality*, Eds. J.C. Bergstrom, K.J. Boyle, and G.L. Poe, Northampton, MA: Elgar, pp. 100–120.

West Virginia Department of Environmental Protection. 2004. WV stream assessment. http://www.dep.state.wv.us/Docs/7714_EPA_WV_2004_IR_Category_StreamListings_appr.pdf (accessed May 2005).

Williams, D.R., M.E. Clark, and J.B. Brown. 1999. *Stream and Water Quality in Coal Mined Areas of the Lower Cheat River Basin, West Virginia and Pennsylvania During Low-Flow Conditions, July 1997.* U.S. Geological Survey Water-Resources Investigations Report 98-4258, U.S. Department of the Interior: Washington, DC.

Williamson, J.M., H.W. Thurston, and M.T. Heberling. 2007. Valuing acid mine drainage remediation options in West Virginia: A preference-calibrated benefit transfer approach. *Environmental Economics and Policy Studies* 8(4): 271–293.

Ziemkiewicz, P., J. Skousen, and J. Simmons. 2003. Long-term performance of passive acid mine drainage treatment systems. *Mine Water and the Environment* 22: 118–129.

6 Economics of Ecosystem Management for the Catawba River Basin

Randall A. Kramer, Jonathan I. Eisen-Hecht, and Gene E. Vaughan

CONTENTS

INTRODUCTION

Ecosystem management is an integrative approach that recognizes the importance of human needs in the wider context of sustaining ecosystems over time. The Ecological Society of America has defined ecosystem management as "driven by explicit goals, executed by policies, protocols, and practices, and made adaptable by monitoring and research based on our best understanding of the ecological interactions and processes necessary to sustain ecosystem composition, structure, and function" (Christensen et al., p. 665). Although the literature is replete with studies of ecosystem management, from a practical standpoint it has proven challenging to integrate natural and social sciences in management settings. For example, studies may focus on ecological aspects of resource management but neglect to incorporate the social and economic information necessary to reach a goal of sustaining the conditions, values, and uses of a resource over time.

Economic information is particularly important for applied studies of ecosystem management. Without appropriate economic information about the value of ecosystem services, it may be difficult for managers to garner financial and political support for protecting resource quality. Also, without such information, it is not possible to make well-informed trade-offs about the social value of maintaining water quality in a river versus the economic value of permitting new industrial or residential development decisions that might degrade water quality. This chapter discusses the importance of nonmarket valuation within the context of ecosystem management and examines how environmental valuation can enhance the understanding of ecosystem functions and values.

In general, the lack of readily available data on the social and economic components of ecosystem management is due to several factors. While the ecological principles that underlie the functioning of ecosystems are relatively well understood, the social and economic aspects of ecosystems are much less clearly defined for the lay and scientific communities (Daily 1997). In addition, much of the traditional research on the values of ecosystems has been isolated within the ecological or economic communities and not readily accessible to specialists in other disciplines (Geoghegan and Bockstael 1997). Metrics abound for the measurement of ecological health (e.g., index of biotic integrity, rapid bioassessment techniques), but measures of economic or social well-being associated with ecosystems or watersheds are not as widely used or understood (Whigham 1997).

Natural resource managers, as they grapple with ecosystem approaches to management, are increasingly realizing the importance of information on the economic valuation of water quality and other ecosystem goods and services (Hanley et al. 1997; Kramer 2007). An important component of ecosystem valuation is the determination of the economic worth of functions and services that do not show up in market transactions.

Nonmarket valuation refers to a variety of tools developed by economists for measuring the value of environmental and other "public" goods that are not traded in existing markets (Pagiola et al. 2004). Due to this public nature, these goods and services are often provided to society at a price that greatly underestimates their true worth. Nonmarket valuation can be utilized to estimate the societal worth of these goods and services and recognizes that this worth is based not only on use values, but also on nonuse values, such as the value of just knowing that a resource exists or is being protected for future generations (Mitchell and Carson 1995). This chapter presents a nonmarket, ecosystem valuation case study for the Catawba River basin in North and South Carolina. The case study integrates a basinwide water quality model with ecosystem valuation methods to estimate the monetary value of managing the Catawba River ecosystem to maintain the current level of water quality in the basin over time.

THE IMPORTANCE OF ECOSYSTEM VALUATION FOR ENVIRONMENTAL RESOURCE MANAGEMENT

An underlying principle of an ecosystem management approach is that human beings are important actors in ecosystem processes. Since virtually all ecosystems have been altered by human activity, an understanding of how human beings relate to these ecosystems is a crucial component of effective and comprehensive ecosystem management. This understanding dictates that effective ecosystem management must address the economic and social aspects of resource issues as well as the ecological ones.

Consideration of the economic aspects of ecosystem services brings to light troubling realizations on how society currently ascribes value to them. Most environmental services are provided to people free of charge or at an arbitrarily determined price that does not adequately reflect their true value. Economists have devised various techniques to arrive at the values of these unpriced ecosystem services and thus at the societal value of goods that provide them (Kramer 2007). One of these is the "contingent valuation" approach, which can measure both use and nonuse values of environmental goods or services and thus their full economic worth (Smith 1997). This approach involves the use of carefully crafted survey instruments that allow people to consider the economic value of the environmental good in question by expressing their willingness to pay (WTP) for changes in the level of provision of the good. The approach avoids simply asking people to imagine that there is a market for the environmental good; rather, the question is asked as a referendum of support for a publicly provided environmental improvement.

WTP, the common metric of stated preference studies, is a widely accepted measure of the satisfaction with a good or resource. It is a measure of the economic compensation needed to keep an individual at the same level of happiness, or utility, after an environmental change as they were before it. WTP can thus be seen as an indication of social well-being and how well-being could improve or decline with proposed changes in the provision of environmental services (Goulder and Kennedy 1997). WTP estimates from a well-designed study can be meaningfully compared to the values of goods and services traded in markets and can provide critically important information for ecosystem managers (Carson et al. 2001).

THE IMPORTANCE OF ECOSYSTEM VALUATION FOR MANAGEMENT OF FRESHWATER RESOURCES

Freshwater is an increasingly important resource as populations expand and the demand for consumptive and nonconsumptive uses of water increases. It is a resource that could greatly benefit from an ecosystem management perspective because of the ecological and social complexities of managing freshwater resources such as lakes and streams. Human demand for freshwater has tripled since 1950 due to population growth, irrigation, and increasing material consumption (Postel and Carpenter 1997). This increased demand for these resources has increased their value; at the same time, these resources have become scarcer. In addition, unlike many other resources, freshwater does not have readily available substitutes. The increasing value of freshwater underlies the importance of bridging the gap between the worth of these resources and their artificially prescribed price, which is often near zero. Unless this true worth is known, these resources are unlikely to be managed in ways that will appropriately protect their quantity and quality for future use. The application of nonmarket ecosystem valuation is thus a crucial step in uncovering this true societal worth.

Estimations of the social worth of resources plays a critical role in ecosystem management as it helps to bridge the social, ecological, and economic dimensions of resource issues. These estimations can then be used to help judge the appropriateness of actions that would affect the provision of a resource in the future. The common scale of measurement for ecosystem valuation studies is the marginal value of a good or how much individuals, or society, would be willing to trade a little bit of something to get a little bit of something else (Toman 1997). This is also the scale at which many environmental decisions are made as policy decisions most commonly involve trade-offs of this nature.

Use of nonmarket valuation methods, however, is not without challenges and controversies. Economists are divided, for instance, on the usefulness and accuracy of stated preference methodologies such as contingent valuation. Some economists claim that the hypothetical nature of these methods casts doubt on their results, while others claim that empirical evidence has shown results of well-conducted stated preference studies to be both valid and reliable (Portney 1994). Additional challenges arise due to the complex nature of many environmental issues and in trying to convey these complexities to survey respondents who may be totally unfamiliar with these issues (Goulder and Kennedy 1997). Many of the concerns about nonmarket valuation methods can be addressed by a very careful survey design process that utilizes focus groups and survey pretesting.

Valuation of ecosystems becomes more challenging as these goods become further removed from market activities. This is particularly true for nonconsumptive uses such as the value placed on the option to use a resource at a later time, or the value placed on knowing resources are protected for their inherent worth or for the use of future generations (Loomis 1996). These nonconsumptive uses of resources may often be their most valuable characteristics. In a study of the preservation of wild and scenic rivers, Sanders et al. (1990) found that nonuse values accounted for about 80% of the total value of the rivers. Other economic valuation studies have found nonuse values to be important when estimating the value of rivers, wetlands, and other water resources (Stevens et al. 1995; Loomis 1996; Bateman et al. 2006).

CASE STUDY: THE CATAWBA RIVER BASIN

The Catawba River basin is located in the Piedmont region of North and South Carolina and encompasses an area of roughly 5,000 square miles. Although sections of the Catawba basin are mostly rural, other sections are experiencing rapid population growth and land use changes. Along its 224-mile course, the Catawba River flows through some of the most populated regions of the Carolinas, including Charlotte, North Carolina. From 2000 to 2006, Charlotte experienced a growth rate of 19.0%, the sixth highest growth rate in U.S. metropolitan areas with populations greater than 1 million people (U.S. Department of Commerce 2007).

The Catawba River is an important and unique feature in the regional landscape and plays a central role in the lives of many residents. The main stem of the river is a series of 11 reservoirs originally created by Duke Power Company for the purpose of hydroelectric power generation. These reservoirs support varied forms of recreation and hydroelectric production as well as cooling water for nuclear and fossil steam power generation and other commercial activities. Many of the surrounding municipal areas get drinking water from the river and return their wastewater to it. Many area residents enjoy the aesthetic beauty of the river from their homes, offices, and businesses. The responsibility for water quality management in the Catawba

River belongs to the states of North and South Carolina. Duke Power Company, in conjunction with regulatory agencies, has taken a supportive role in monitoring the health of the river. Water quality monitoring data collected by all agencies show a disturbing trend of decreasing water quality once the river flows past Charlotte (South Carolina Department of Health and Environmental Control, 2005). Rapid population growth and land use changes contribute to these water quality problems and potentially threaten the health and vitality of the Catawba River.

In examining the potential for ecosystem management in the Catawba basin, Duke Power Company recognized the importance of conducting a nonmarket economic valuation of the Catawba River basin. With this information, all stakeholders would, perhaps for the first time, have information on the economic and social value of this resource and a tool to monitor that change in value over time. Stakeholders could also use this information to weigh the costs and benefits of actions that could have an impact on the water quality of the Catawba River and thus its value to the surrounding communities (Christman and Wayman 1996).

THE ROLE OF ECOSYSTEM MANAGEMENT FOR THE CATAWBA RIVER

The management of watersheds can provide tremendous challenges to resource managers. Watersheds cross political boundaries, and effective management often necessitates communication among several different local, state, and federal agencies. The Catawba River basin crosses two states as well as 14 counties and many cities. Duke Power Company is unique among stakeholders to this resource in that they own land and have power-generating facilities throughout the basin and thus have a broader geographical focus than many other stakeholders. In recognition of the difficulty of managing the watershed by traditional institutions and concepts, Duke Power has worked with other stakeholders to promote an ecosystem management paradigm for the Catawba River. In an initial step toward an ecosystem management approach, Duke Power's Environmental Laboratory has conducted studies of the ecosystem characteristics of the watershed, including its water chemistry, aquatic ecology, geography, wildlife, land use, demographics, and development history. The company has identified key steps necessary for the development of a successful ecosystem management effort (Christman and Wayman 1996). One of these steps is the development of information about the social worth of the Catawba River's water quality. With this information, stakeholders would have better information about the value of the resource and how that value might change over time.

One crucial step identified by the company was the formation of local partnerships specific to the Catawba basin. A diverse array of partnerships would help to ensure an appropriate level of concern for the resource and would facilitate in the sharing of information with other stakeholders and the public. Partnerships are also useful in helping stakeholders to identify the most important issues and in deciding on criteria for evaluating proposed projects that would affect the river. Partnerships could also help to identify current gaps in information (Christman and Wayman 1996). In 1992, the Bi State Catawba River Task Force was formed to raise awareness of resource issues and to provide a forum for management discussions.

The Economic Value of Water Quality in the Catawba River Basin

In 1997, Duke Power Company funded a study at the Nicholas School of the Environment at Duke University to conduct a nonmarket valuation of water quality in the Catawba River basin (Kramer and Eisen-Hecht 2002). The objective of the study was to estimate the economic value of water quality in the basin and the economic benefits of protecting the current level of water quality over time in the face of rapid population growth in the region. This 2-year study was completed in 1999.

This study used the contingent valuation method (CVM) to assess the WTP of Catawba River basin residents for a management plan that was designed to protect the current level of water quality over time. This approach allowed direct queries to individuals through a survey about their WTP to avoid future declines in river water quality. The survey instrument was developed over the first year of the study. Design of the survey involved an extensive literature review, meetings with various stakeholder groups, focus groups conducted with area residents, and an extensive pretest of the survey using both in-person and telephone interview formats. The survey process was carefully designed to ensure a representative sample of area residents, to provide a careful description of water quality issues and the way in which a river basin management plan could protect water quality, and to use state-of-the-art methods to obtain reliable estimates of WTP.

Survey data collection occurred between September and December 1998. Surveys were administered through a combined mail and telephone format, with all questions answered by telephone. Survey respondents were sent by mail an information booklet, "Water Quality in the Catawba River Basin," which described the relevant issues and included color photos and maps of the Catawba River basin. Hagler Bailly, a leading market research firm, conducted the telephone interviews, which averaged 24 minutes in length. In total, 1,085 households completed the survey. These households were sampled at random from the 16 counties with 10 or more square miles in the Catawba River basin (11 of these counties are in North Carolina, and 5 are in South Carolina). The overall response rate of the survey was 47%, which is within the range of response rates for surveys conducted with similar approaches (Smith et al. 1997; Smith and Mansfield 1998; Blomquist et al. 2003).

The survey respondents were fairly evenly split by gender and had an average age of 50 (see Table 6.1). The median level of educational achievement was some college, and the median annual household income of the sample was $45,000 to $60,000. The sample was somewhat more educated and wealthier than the U.S. Census data reported for the study area. Weighting techniques were used to correct for these differences in the estimation of WTP reported in this chapter.

Survey respondents answered a variety of questions about their use of the Catawba River, their perceptions of area water quality, and their socioeconomic profiles. Responses to these questions underscored the importance of the Catawba River to area residents. Of the survey respondents, 57% had heard of water quality concerns in the Catawba River basin prior to taking this survey. In addition, 39% of the respondents said protecting area water quality was more important than other environmental issues in their state, and 59% said that this issue was just as important as other issues. Forty-nine percent of the sample thought that water quality in their

TABLE 6.1
Descriptive Statistics for Survey Sample

Socioeconomic Characteristic	Percentage or Mean/Median Value (Survey Data)	Percentage or Mean/Median Value (Census Data)
Percentage female	46% ($n = 1,085$)	52%
Mean age	50 ($n = 1,070$)	49
Percentage high school graduates	92% ($n = 1,082$)	69%
Percentage college graduates	39% ($n = 1,082$)	17%
Percentage Caucasian	89% ($n = 1,072$)	80%
Percentage African American	7% ($n = 1,072$)	19%
Mean annual household income	$55,481 ($n = 989$)	$45,477

Source: Kramer and Eisen-Hecht, 2002.

area had gotten worse over the last 5 years, and only 8% thought it had improved in that time.

The central element of the survey involved the contingent valuation scenario in which respondents were asked to place a value on the protection of area water quality. Several steps were involved in the presentation of this scenario to respondents. Color maps included in the survey information booklet showed area water quality, depicted as good, fair, or poor, and how it could change over time given projected population and land use changes. The first map presented the status quo of water quality in the basin based on information obtained from the water regulatory agencies in North and South Carolina. The second map was developed using the watershed analysis risk management framework (WARMF) model (Chen et al. 1998) and showed a possible future scenario of what area water quality could be like in 10 years if the resulting population growth and land use changes were not actively managed.

With these maps, respondents were also presented with a potential management plan for water quality in the basin (see Box 6.1). Developed in consultation with state regulatory agencies, this plan was believed to be adequate for the protection of water quality at its current level, or the level shown in the first map, over time. This plan consisted of four main strategies: use of best management practices in farms, construction sites, and residential areas; development of a basinwide land use plan; upgrading and improving area wastewater treatment plants; and purchasing and setting aside critical tracts of land. Respondents were told that, given the available information, it was likely that this plan would be successful in protecting the current level of water quality in the basin over time.

Respondents were then asked the contingent valuation question, which is shown in Box 6.2. In accordance with standard practice methods of conducting the CVM, this question was proposed to respondents as a referendum on which they could vote. This referendum was offered to them at one of eight different prices, ranging from $5 to $250 per year for the next 5 years. These different values were assigned randomly

Box 6.1 Summary of Water Quality Management Plan Presented to Catawba Basin Survey Respondents

This management plan addresses the main water pollution problems in the basin: sediment and nutrients. It also continues to manage related problems such as pollution by toxic substances, bacteria, and viruses. While this specific management plan has not been proposed by state governmental agencies, it is drawn from their best available information. This includes information on the condition of the basin and how to best manage the problems.

This potential management plan includes the following components:

1. Construction and use of best management practices (BMPs) within the basin. These include buffer strips and holding ponds for farms, construction sites, and residential areas.
2. Development of a basinwide land use plan. This would encourage land uses in the basin that are consistent with the goals for water quality in the basin. Government agencies could use this land use plan to make decisions that would affect water quality.
3. Improving and increasing the capacity of sewage treatment plants in cities within the basin.
4. Purchasing and setting aside tracts of land that have been determined as critical to the protection of water quality.

Box 6.2 The Contingent Valuation Question for Valuing the Water Quality Management Plan

Now, assume a vote is being held today to approve or reject this management plan. Your payment for this plan would be collected through an increase in your usual state income taxes. All residents in counties within the Catawba River basin would make identical payments. This money would *only* be used for implementing this management plan for the Catawba River basin. If a majority of Catawba River basin county residents vote in favor of this management plan, it will go into effect. Before you answer the following question, please consider your current income as well as your expenses:

Suppose that this management plan would cost you $____ (5, 10, 25, 50, 100, 150, 200, 250) each year for the next 5 years in increased state income taxes. Would you vote in favor of the management plan?

to respondents by computer before any additional information was collected about them, such as their income or their use of the Catawba River.

Sixty-six percent of the respondents said that they would vote for the management plan at the various prices at which it was offered to them, and 31% said they would not support the plan. An even larger proportion of the respondents, 76%, thought the management plan was likely to succeed in protecting area water quality. An econometric analysis (see Glossary for definition) was used to examine the relationship between support for the management plan and survey respondents' characteristics. Results of the econometric analysis showed that support of the management plan was positively correlated with respondents' income, education level, support of environmental organizations, and perceived importance of protecting water quality both for use of the river and for its existence value (Kramer and Eisen-Hecht 2002). Support of the management plan was also correlated with respondents' state of residence, with downstream residents in South Carolina more willing to pay for the plan than upstream residents in North Carolina, who experience fewer water quality problems. All of these statistical results support the hypothesized effects of the explanatory variables.

As economic theory would predict, support of the management plan was negatively correlated with the price at which the plan was offered to respondents. As shown in Figure 6.1, the percentage of respondents willing to support the plan declined steadily as the price of the plan increased. Of the respondents who were offered the plan at $5 per year, 88% voted in favor of the plan. At the $250 price level, the number supporting the plan was cut by more than half to 41%. These results indicate that, like private goods, water quality in the Catawba River basin has a downward sloping demand curve, and as the price for it goes up, the demand for it goes down.

After the CV (contingent valuation) question, the survey contained various questions designed to elicit additional information from respondents regarding their votes on the management plan. One of these questions sought to uncover the most important reasons why respondents would value the management plan. Respondents were asked to rate the importance of different reasons why the management plan might be of value to them (Table 6.2). The quality of area drinking water received the highest

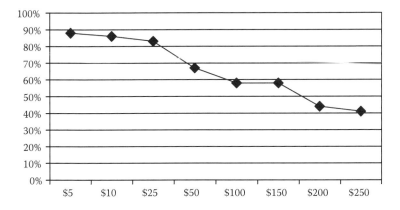

FIGURE 6.1 Percentage of respondents voting for the management plan at each price level.

TABLE 6.2
**Respondents' Ratings of the Importance of Reasons Why They Would Value
a Management Plan for Water Quality in the Catawba River Basin ($n = 1,085$)[a]**

Reason	Mean
Drinking water quality in the area	4.77
Just knowing the river is being protected, regardless of respondents' use of it	4.25
Use of the river by respondents' friends and family	3.74
Respondents' own recreational use of the river	3.52

[a] Reasons were rated on a scale of 1 to 5 with 1 not being important at all and 5 being very important.

rating, followed by the knowledge that the waters in the basin were being protected, regardless of respondents' use of them. These results clearly show that support for water quality protection is motivated by more than direct use considerations, adding further justification to the use of an ecosystem management approach to managing the river basin.

In CV studies, a mean WTP value is typically estimated from individual responses to the CV question. Through a procedure developed by Turnbull (1976) and used in several other CV studies (Carson et al. 1994; Garrod and Willis 1999), the mean WTP was estimated to be $139 per year per Catawba River basin taxpayer. This WTP is equal to the well-being that people receive from knowing that water quality is being protected at its current level over time. Table 6.3 shows a distribution of WTP values. Respondents in downstream South Carolina were willing to pay more for water quality protection. In both states, WTP increased with average income levels.

TABLE 6.3
Willingness to Pay for Water Quality Protection in the Catawba River

Respondent Group	Mean Willingness to Pay
Total Sample	$139
Comparison across states	
North Carolina residents	$135
South Carolina residents	$150
Comparison across income levels	
Household income $30,000 and under	$116
Household income between $30,001 and $75,000	$157
Household income above $75,000	$180

The $139 mean WTP value was aggregated to obtain a measure of the total economic benefits accruing to area residents from the protection of Catawba basin water quality. Aggregation involved correcting the sample for differences in the demographics of the sample and the general population (as reported by 1990 Census data). The mean WTP value was then multiplied by an estimate of state income taxpayers in the 16-county area of the sample to obtain a total annual economic benefit of $75.4 million resulting from the implementation of a management plan to protect area water quality at its current level over time.

CONCLUSION

Results of the economic valuation study of the Catawba River have important consequences for river stakeholders. Knowledge of the economic magnitude of protecting water quality is a critical step in weighing the trade-offs associated with actions that could affect the health of this ecosystem. Converting riparian zones into housing developments, adding a new wastewater treatment plant along a lake, or transporting drinking water from one basin to another all have environmental consequences that can be measured through scientific methods. The question of whether these actions are desirable, however, are societal decisions that are most often made by evaluating the proposed costs of various projects. Efforts like this study enable policy makers, regulators, and the general citizenry to make more informed decisions regarding ecosystem resources as they shed light on the societal costs and benefits associated with their use or preservation.

REFERENCES

Bateman, I.J., B.H. Day, S. Georgiou, and I. Lake. 2006. The aggregation of environmental benefit values: Welfare measures, distance decay and total WTP. *Ecological Economics* 60: 450–460.

Blomquist, G.C., M.A. Newsome, and D.B. Stone. 2003. Measuring principals' values for environmental budget management: An exploratory study. *Journal of Environmental Management* 68: 83–93.

Carson, R.T., W.M. Hanemann, R.J. Kopp, J.A. Krosnick, R.C. Mitchell, S. Presser, P.A. Rudd, and V.K. Smith. 1994. *Prospective Interim Lost Use Value Due to DDT and PCB Contamination in the Southern California Bight: Volume 1*. Report to the National Oceanic and Atmospheric Administration, Natural Resource Damage Assessment: LaJolla, CA.

Carson, R.T., N.E. Flores, and N.F. Meade. 2001. Contingent valuation: Controversies and evidence, *Environmental and Resource Economics* 19: 173–210.

Chen, C.W., J. Herr, and L. Ziemelis. 1998. *Watershed Analysis Risk Management Framework—A Decision Support System for Watershed Approach and TMDL Calculation*. Documentation Report TR110809. Palo Alto, CA: Electric Power Research Institute.

Christensen, N.L., A.M. Bartuska, J.H. Brown, S. Carpenter, C. D'Antonio, R. Francis, J.F. Franklin, J.A. MacMahon, R.F. Noss, D.J. Parsons, C.H. Peterson, M.G. Turner, and R.G. Woodmansee. 1996. The Report of the Ecological Society of America Committee on the Scientific Basis for Ecosystem Management. *Ecological Applications* 6(3): 665–691.

Christman, J.D., and A.D. Wayman. 1996. The ecosystem approach: A management alternative for Duke Power Company and the Catawba River Basin. Master's project, Duke University, Durham, NC.

Daily, G.C. 1997. Introduction: What are ecosystem services? in *Nature's Services: Societal Dependence on Natural Ecosystems*, G.C. Daily, Ed. Washington, DC: Island Press, pp. 1–10.

Garrod, G., and K.G. Willis. 1999. *Economic Valuation of the Environment*. Northampton, MA: Elgar.

Geoghegan, J., and N. Bockstael. 1997. Human behavior and ecosystem valuation: An application to the Patuxent Watershed of the Chesapeake Bay, in *Ecosystem Function and Human Activities*, R.D. Simpson and N.L. Christensen Jr., Eds. New York: Chapman and Hall, pp. 147–174.

Goulder, L.H., and D. Kennedy. 1997. Valuing ecosystem services: Philosophical bases and empirical methods, in *Nature's Services: Societal Dependence on Natural Ecosystems*, G.C. Daily, Ed. Washington, DC: Island Press, pp. 23–48.

Hanley, N., J.F. Shogren, and B. White. 1997. *Environmental Economics in Theory and Practice*. New York: Oxford University Press.

Kramer, R.A. 2007. Economic valuation of ecosystem services, in *Sage Handbook on Environment and Society*, J. Pretty, T. Benton, J. Guivant, D.R. Lee, D. Or, and M. Pfeffer, Eds. Thousand Oaks, CA: Sage, pp. 172–180.

Kramer, R.A., and J.I. Eisen-Hecht. 2002. Estimating the economic value of water quality in the Catawba River Basin. *Water Resources Research* 38(9): 1–10.

Loomis, J.B. 1996. Measuring the economic benefits of removing dams and restoring the Elwha River: Results of a contingent valuation survey. *Water Resources Research* 32(2): 441–447.

Mitchell, R.C., and R.T. Carson. 1995. Current issues in the design, administration, and analysis of contingent valuation surveys, in *Current Issues in Environmental Economics*, P. Johansson, B. Kristrom, and K. Maler, Eds. Manchester, U.K.: Manchester University Press, pp. 10–34.

Pagiola, S., K. von Ritter, and J. Bishop. 2004. *Assessing the Value of Ecosystem Conservation*. World Bank Environmental Department Paper no. 101. The World Bank Environmental Department, in Collaboration with the Nature Conservancy and IUCN—The World Conservation Union, Washington, DC.

Postel, S., and S. Carpenter, 1997. Freshwater ecosystem services, in *Nature's Services: Societal Dependence on Natural Ecosystems*, G.C. Daily, Ed. Washington, DC: Island Press, pp. 195–214.

Portney, P.R. 1994. The contingent valuation debate: 'Why economists should care.' *Journal of Economic Perspectives* 8(4): 3–17.

Sanders, L.D., R.G. Walsh, and J.B. Loomis. 1990. Toward empirical estimation of the total value of protecting rivers. *Water Resources Research* 26(7): 1345–1357.

Smith, V.K. 1997. Pricing what is priceless: A status report on non-market valuation of environmental resources, in *The International Yearbook of Environmental and Resource Economics 1997/1998*, H. Folmer and T. Tietenberg, Eds. Cheltenham, U.K.: Elgar.

Smith, V.K., X. Zhang, and R.B. Palmquist. 1997. Marine debris, beach quality, and non-market values. *Environmental and Resource Economics* 10: 223–247.

Smith, V.K., and C. Mansfield. 1998. Buying time: Real and hypothetical offers. *Journal of Environmental Economics and Management* 36(3): 209–224.

South Carolina Department of Health and Environmental Control (SCDHEC). 2005. *Watershed Water Quality Assessment: Catawba River Basin*. Technical Report no. 012-05. Columbia, SC: Bureau of Water.

Stevens, T.H., S. Benin, and J.S. Larson. 1995. Public attitudes and economic values for wetland preservation in New England. *Wetlands* 15(3): 226–231.

Toman, M.A. 1997. Ecosystem valuation: An overview of issues and uncertainties, in *Ecosystem Function and Human Activities*, R.D. Simpson and N.L. Christensen Jr., Eds. New York: Chapman and Hall, pp. 25–44.

Turnbull, B. 1976. The empirical distribution function with arbitrary grouped, censored, and truncated data. *Journal of the Royal Statistical Society* B, 38: 290–295.

U.S. Department of Commerce. 2007. *50 Fastest-Growing Metro Areas Concentrated in West and South*. U.S Census Bureau release of April 5. CB07-51, U.S. Census Bureau News, Washington, DC.

Whigham, D.F. 1997. Ecosystem functions and ecosystem values, in *Ecosystem Function and Human Activities*, R.D. Simpson and N.L. Christensen Jr., Eds. New York: Chapman and Hall, pp. 225–240.

7 Estimating Willingness to Pay for Aquatic Resource Improvements Using Benefits Transfer

Robert J. Johnston and Elena Y. Besedin

CONTENTS

INTRODUCTION

As we learned in Chapter 5, benefits transfer may be characterized as the "practice of taking and adapting value estimates from past research ... and using them ... to assess the value of a similar, but separate, change in a different resource"

(Smith et al. 2002, p. 134). It involves adapting research conducted for another purpose to estimate values within a particular policy context (Bergstrom and De Civita 1999). Although primary research methods are generally considered to be superior to benefits transfer, resource or data constraints often render benefits transfer the only viable means to estimate nonmarket values. Benefits transfer methods may be placed into three general categories: (1) transfer of an unadjusted fixed-value estimate generated from a single study; (2) the use of expert judgment to aggregate or otherwise alter benefits to be transferred from a site or set of sites; and (3) estimation of a value estimator model or *benefits transfer function*, often based on data gathered from multiple sites (Bergstrom and De Civita 1999). Given the generally unreliable performance of unadjusted single-site transfers, researchers are increasingly considering approaches that allow welfare measures to be adjusted for characteristics of the policy context using the last two methods (Johnston et al. 2005; U.S. Environmental Protection Agency [USEPA] 2000).

This chapter is the most technical in this book. We recommend this chapter for those who are quite comfortable with the material presented thus far, and who want to delve deep into the valuation procedures. However, it does provide a basic understanding of benefits transfer and highlights some of the challenges of using this method.

When researchers have access to a large number of prior studies estimating values for a particular natural resource in other locations or policy contexts, it is possible to estimate benefit functions using statistical analysis that synthesizes and combines findings from these studies. This method is called *meta-analysis*. Glass (1976, p. 3) characterized meta-analysis as "the statistical analysis of a large collection of results for individual studies for the purposes of integrating the findings." When used for benefits transfer, meta-analysis assumes the existence of "an underlying meta-valuation function that relates the magnitude of empirical estimates of value to characteristics of the study site, market and research methods" (Rosenberger and Stanley 2006, p. 373). The estimated valuation or benefit function allows researchers to more appropriately adjust willingness-to-pay (WTP) estimates, providing an improved mechanism for benefits transfer (Rosenberger and Loomis 2003). This allows for the estimation of benefit functions that are often better able to forecast values in a wide range of policy contexts, thereby providing more valid benefit estimates for transfer applications. Based on this potential, USEPA (2000, p. 87) guidelines characterized meta-analysis as "the most rigorous benefits transfer method."[1]

This chapter describes the use of meta-analysis for applied, function-based benefits transfer and illustrates an application to aquatic habitat improvements.

Although the estimation of original meta-analysis models requires a fair degree of expertise, the use of already-estimated models for benefits transfer (i.e., to estimate non-market values for appropriate policy contexts) requires only basic mathematical skills. Hence, this chapter emphasizes how one would use high-quality meta-analyses that have already been estimated to quantify benefits for specific policy changes. Those interested in the technical details involved in the estimation of original meta-analyses are directed to the Appendix as well as to sources such as Johnston et al. (2005; 2006b), Bateman and Jones (2003), and Rosenberger and Loomis (2000a,b).

BENEFIT FUNCTIONS AND META-ANALYSIS

Theory tells us that nonmarket values for changes in the quality or quantity of natural resources should vary according to variables that characterize the resource, policy context, and affected populations (Bergstrom and Taylor 2006). Based on theory and findings from past research, we expect various attributes to be associated with systematic variations in nonmarket values (e.g., larger environmental improvements should, on average, be associated with higher nonmarket values[2]). As a result, these attributes may be used to forecast nonmarket values in a variety of policy contexts (Johnston et al. 2005). Benefit functions are mathematical functions that allow researchers to predict nonmarket values for particular policy settings and resource changes based on context-specific values for variables characterizing such features as (1) affected populations, (2) geographic region and size, (3) natural resource attributes, (4) baseline resource condition and the magnitude of change, and (5) other attributes as guided by theory and past findings. The structure and content of benefit functions should follow basic economic theory (Bergstrom and Taylor 2006). Once the benefit function has been estimated, the analyst seeking to estimate nonmarket value (often WTP estimates) for a particular site and policy context "plugs in" values for these variables. These values, or variable levels, are chosen to best fit the attributes of the site, resource, and policy in question. Given these selected variable values, the benefit function provides an estimate of nonmarket value tailored to the particular site, resource, and policy context.

The following sections describe the estimation of a benefit function using meta-analysis and illustrate the use of the function to conduct benefits transfer. To illustrate the methods involved, we draw from a case study involving WTP for water quality improvements in aquatic habitats, conducted originally to explore WTP for fish and related resources affected by USEPA regulations (Johnston et al. 2005). Although we summarize the statistical methods used to estimate the original meta-analysis, primary emphasis is given to the use of meta-analysis results to conduct function-based benefits transfer. We also highlight potential challenges in the design and use of meta-models and associated benefit functions for policy analysis.

CASE STUDY DATA AND CONCEPTUAL APPROACH

The goal of this case study was to estimate a benefit function that captures the relative influence of resource, context, and study characteristics on per household WTP for water quality improvements that affect aquatic species, based on patterns observed in prior research studies. Given this emphasis, data for the meta-analysis were drawn from a large set of prior nonmarket valuation studies that estimated total WTP for water quality changes that affect aquatic habitats. From more than 300 identified surface water valuation studies addressing water quality changes, 34 were found to be suitable for inclusion in the meta-data. We chose these 34 because (1) the study estimates total (use and nonuse) per household WTP, (2) the water quality change being valued affects aquatic habitat in a water body that provides recreational fishing uses or other recreational activities, (3) the study was conducted in the United States, (4) the study applies research methods generally accepted by journal literature, and (5) the study provides sufficient information regarding resource, context, and study attributes to allow inclusion in the meta-data.

The resulting meta-data comprise 81 observations from 34 unique studies conducted between 1973 and 2001 (Johnston et al. 2005). There are more observations than studies because many studies provided more than one estimate of WTP. Multiple WTP estimates from single studies were available due to in-study variations in such factors as the extent of amenity change, elicitation methods applied, water body type and number, recreational activities affected, and species affected. Due to the requirement that each study estimate total (use and nonuse) WTP, the data were limited to studies relying on stated preference methods; these included contingent valuation, stated choice model approaches, and combined revealed/stated preference techniques. Table 7.1 summarizes principal study characteristics for those studies included in the meta-data.

Table 7.2 summarizes the set of independent variables included in the meta-analysis. These are the variables that were expected to explain observed patterns in WTP.[3] For ease of exposition, these variables are categorized into those characterizing (1) study and methodology, (2) surveyed populations, (3) geographic region and scale, (4) water body type, and (5) resource condition and change. *Study and methodology variables* included elements such as the year a study was conducted; payment vehicle (e.g., whether individuals were told in the survey that their payments were voluntary or mandatory or if payments would occur on an annual or lump sum basis) and elicitation format (e.g., whether in-person interviews or mail surveys were used); WTP estimation methods and conventions; and survey response rates. *Surveyed population variables* included, for example, the average income of respondents and the representation of users and nonusers within the survey sample. *Geographic region and scale* variables characterized features such as the number of water bodies affected by the policy and the geographic region in which the study was

TABLE 7.1
Characteristics of Surface Water Valuation Studies Included in Meta-Analysis

Citation for Study	Number of Observations in Meta-Data	State	Water Body Type	Species Affected	Methodology	Adjusted Raw WTP Values[a]
Aiken (1985)	1	CO	All freshwater	Game fish	CV—multiple methods[b]	$167.98
Anderson and Edwards (1986)	1	RI	Salt pond/ marshes	Unspecified	CV—open ended	$157.14
Azevedo et al. (2001)	5	IA	Lake	Game fish	CV—discrete choice	$17.76– $118.68
Bockstael et al. (1989)	2	MD	Estuary	Unspecified	CV—discrete choice	$65.80– $209.51
Cameron and Huppert (1989)	1	CA	River/ stream	Game fish	CV—discrete choice	$43.07
Carson et al. (1994)	2	CA	Estuary	Game fish; multiple categories	CV—discrete choice	$35.83– $67.47
Clonts and Malone (1990)	3	AL	River/ stream	Unspecified	CV—iterative bidding	$68.10– $110.85
Croke et al. (1987)	9	IL	River/ stream	All recreational fish; none	CV—iterative bidding	$53.31– $81.46
Cronin (1982)	4	DC	River/ stream	All recreational fish	CV—open ended	$61.85– $212.73
Desvousges et al. (1983)	2	PA	River/ stream	Unspecified	CV—discrete choice	$111.41– $220.24
De Zoysa (1995)	2	OH	Lake; river and lake	Multiple categories	CV—discrete choice	$35.88– $61.02
Farber and Griner (2000)	3	PA	River/ stream	All recreational fish	CV—discrete choice	$44.22– $105.58
Hayes et al. (1992)	2	RI	Estuary	Shellfish; none	CV—discrete choice	$339.72– $351.47
Herriges and Shogren (1996)	2	IA	Lake	All recreational fish	CV—discrete choice	$53.66– $180.35
Huang et al. (1997)	2	NC	Estuary	Multiple categories	CV—discrete choice; revealed and stated preference	$221.75– $228.07
Kaoru (1993)	1	MA	Salt pond/ marshes	Shellfish	CV—open ended	$190.10
Lant and Roberts (1990)	3	IA/IL	River/ stream	Game fish; all recreational fish	CV—discrete choice	$107.86– $134.18

(continued)

TABLE 7.1 (continued)

Citation for Study	Number of Observations in Meta-Data	State	Water Body Type	Species Affected	Methodology	Adjusted Raw WTP Values[a]
Loomis (1996)	1	WA	River/ stream	Game fish	CV—discrete choice	$80.93
Lyke (1993)	2	WI	Lake	Game fish	CV—discrete choice	$51.96– $84.99
Magat et al. (2000)	2	CO/NC	All freshwater	All aquatic species	CV—iterative bidding	$114.49– $376.61
Matthews et al. (1999)	2	MN	River/ stream	All aquatic species	CV—discrete choice	$15.77– $22.01
Mitchell and Carson (1981)	1	National	All freshwater	All aquatic species	CV—discrete choice	$242.34
Olsen et al. (1991)	3	Pacific NW	River/ stream	Game fish	CV—open ended	$34.48– $107.59
Roberts and Leitch (1997)	1	MN/SD	Lake	Multiple categories	CV—discrete choice	$7.26
Rowe et al. (1985)	1	CO	River/ stream	Game fish	CV—open ended	$117.04
Sanders et al. (1990)	4	CO	River/ stream	Unspecified	CV—open ended	$70.44– $171.59
Schulze et al. (1995)	2	MT	River and lake	Multiple categories	CV—discrete choice	$15.08– $21.16
Stumborg et al. (2001)	2	WI	Lake	Multiple categories	CV—discrete choice	$57.90– $88.38
Sutherland and Walsh (1985)	1	MT	River and lake	Unspecified	CV—open ended	$126.98
Welle (1986)	6	MN	All freshwater	Multiple categories; game fish	Multiple methods	$95.30– $207.32
Wey (1990)	2	RI	Salt pond/ marshes	Shellfish	Multiple methods	$55.61– $200.50
Whitehead and Groothuis (1992)	3	NC	River/ stream	All recreational fish	CV—open ended	$27.74– $46.23
Whitehead et al. (1995)	2	NC	Estuary	Multiple categories	CV—iterative bidding	$68.08– $97.91
Whittington et al. (1994)	1	TX	Estuary	All aquatic species	CV—discrete choice	$169.32

CV, contingent valuation.

[a] As noted in the text, reported WTP values apply to different levels of water quality change. All WTP estimates are converted to 2002 dollars and rounded to the nearest cent and hence may not match exactly the raw WTP estimates reported in source studies. If multiple WTP estimates were available from a given study, the range of values is presented.

[b] The author averaged WTP estimates derived from both open-ended and iterative bidding methods to obtain a single reported WTP estimate.

TABLE 7.2
Meta-Analysis Variables and Descriptive Statistics

Variable	Description	Units and Measurement	Mean (SD)
ln_WTP	Natural log of willingness to pay for specified resource improvements. WTP for all studies was converted to 2002 dollars using the U.S. Bureau of Labor Statistics nonseasonally adjusted average CPI for all urban consumers.	Natural log of dollars (range: 1.98 to 5.93)	4.43 (0.77)
year_indx	Year in which the study was conducted, converted to an index by subtracting 1970.	Year index (range: 3 to 31)	18.79 (6.57)
discrete_ch	Binary variable indicating that WTP was estimated using a discrete choice survey instrument.	Binary (range: 0 or 1)	0.35 (0.37)
voluntary	Binary variable indicating that WTP was estimated using a payment vehicle described as voluntary.	Binary (range: 0 or 1)	0.07 (0.26)
interview	Binary variable indicating that the survey was conducted using in-person interviews.	Binary (range: 0 or 1)	0.19 (0.39)
mail	Binary variable indicating that the survey was conducted through the mail.	Binary (range: 0 or 1)	0.56 (0.50)
lump_sum	Binary variable indicating that payments were to occur on something other than a long-term annual basis (e.g., a single lump sum payment).	Binary (range: 0 or 1)	0.21 (0.41)
nonparam	Binary variable indicating that WTP was estimated using nonparametric methods.	Binary (range: 0 or 1)	0.46 (0.50)
wq_change	Change in mean water quality, specified on the RFF water quality ladder. Defined as the difference between baseline and postimprovement quality. Where the original study (survey) did not use the RFF water quality ladder, we mapped water quality descriptions to analogous levels on the RFF ladder to derive water quality change (see text). Note that this variable was only included in the model as part of an interaction term (*WQ_fish, WQ_shell, WQ_many, WQ_non*).	Water quality ladder units (range: 0.5 to 5.75)	2.42 (1.07)
lnwq_change	The natural log of *wq_change* (see above).	Range: −0.69 to 1.75	0.77 (0.52)
wq_ladder	Binary variable indicating that the original survey reported resource changes using a standard Resources for the Future water quality ladder.	Binary (range: 0 or 1)	0.32 (0.47)

(*continued*)

TABLE 7.2 (continued)

Variable	Description	Units and Measurement	Mean (SD)
protest_bids	Binary variable indicating that protest bids were excluded when estimating WTP.	Binary (range: 0 or 1)	0.46 (0.50)
outlier_bids	Binary variable indicating that outlier bids were excluded when estimating WTP.	Binary (range: 0 or 1)	0.22 (0.42)
median_WTP	Binary variable indicating that the study reported median, not mean, WTP.	Binary (range: 0 or 1)	0.06 (0.24)
hi_response	Binary variable indicating that the survey response rate exceeds 74% (i.e., 75% or above).	Binary (range: 0 or 1)	0.31 (0.47)
income	Mean income of survey respondents, either as reported by the original survey or calculated based on U.S. Census averages for the original surveyed region.	Dollars (range: 30,396 to 137,693)	47,034.10 (12,788.72)
nonusers	Binary variable indicating that the survey is implemented over a population of nonusers (default category for this dummy is a survey of any population that includes users).	Binary (range: 0 or 1)	0.19 (0.39)
single_river	Binary variable indicating that resource change explicitly takes place over a single river (default is a change in an estuary).	Binary (range: 0 or 1)	0.24 (0.43)
single_lake	Binary variable indicating that resource change explicitly takes place over a single lake.	Binary (range: 0 or 1)	0.12 (0.33)
multiple_river	Binary variable indicating that resource change explicitly takes place over multiple rivers.	Binary (range: 0 or 1)	0.09 (0.28)
salt_pond	Binary variable indicating that resource change explicitly takes place over multiple salt ponds.	Binary (range: 0 or 1)	0.05 (0.22)
num_riv_pond	Number of rivers or salt ponds affected by policy when *multiple_river* or *salt_pond* = 1. (Only studies addressing rivers and salt ponds specified multiple water bodies.) Specified as the sum of the multiplicative interactions between *multiple_river* and the number of water bodies and that of *salt_pond* and the number of water bodies.	Number of specified rivers or ponds (range: 0 to 15)	1.40 (3.56)
regional_fresh	Binary variable indicating that resource change explicitly takes place in a fresh water body.	Binary (range: 0 or 1)	0.16 (0.37)
southeast	Binary variable indicating that survey was conducted in the USDA Southeast region (default is Northeast region).	Binary (range: 0 or 1)	0.12 (0.33)

TABLE 7.2 (continued)

Variable	Description	Units and Measurement	Mean (SD)
southeast	Binary variable indicating that survey was conducted in the USDA Southeast region (default is Northeast region).	Binary (range: 0 or 1)	0.12 (0.33)
pacif_mount	Binary variable indicating that survey was conducted in the USDA Pacific/Mountain region.	Binary (range: 0 or 1)	0.18 (0.40)
plains	Binary variable indicating that survey was conducted in the USDA Northern or Southern Plains region.	Binary (range: 0 or 1)	0.02 (0.15)
mult_reg	Binary variable indicating that survey included respondents from more than one of the regions.	Binary (range: 0 or 1)	0.04 (0.19)
WQ_fish	Interaction variable: *wq_change* multiplied by a binary variable identifying studies in which water quality improvements are stated to benefit only fin fish. Default is zero (i.e., change did not affect fish).	Water quality ladder units (range: 0.5 to 5.75)	1.15 (1.53)
WQ_shell	Interaction variable: *wq_change* multiplied by a binary variable identifying studies in which water quality improvements are stated to benefit only shellfish. Default is zero (i.e., change did not affect shellfish).	Water quality ladder units (range: 0.5 to 4.00)	0.12 (0.64)
WQ_many	Interaction variable: *wq_change* multiplied by a binary variable identifying studies in which water quality improvements are stated to benefit multiple species types. Default is zero (i.e., change did not affect multiple species).	Water quality ladder units (range: 0.5 to 4.00)	0.63 (1.20)
WQ_non	Interaction variable: *wq_change* multiplied by a binary variable identifying studies in which species benefiting from water quality improvements remain unspecified. Default is zero (i.e., change did not affect unspecified species).	Water quality ladder units (range: 0.5 to 2.5)	0.52 (0.93)
nonfish_uses	Binary variable identifying studies in which changes in uses other than fishing are specifically noted in the survey.	Binary (range: 0 or 1)	0.73 (0.45)
fishplus	Binary variable identifying studies in which a fish population or harvest change of 50% or greater is reported in the survey.	Binary (range: 0 or 1)	0.12 (0.33)
baseline	Baseline water quality, specified on the RFF water quality ladder.	Water quality ladder units (range: 0 to 7)	4.60 (2.47)

SD, standard deviation.

conducted. *Water body type* variables included hydrological characteristics of the affected water body (e.g., river, lake, salt pond, estuary). Finally, *resource condition and change* variables characterized baseline conditions, resource uses supported, and the extent of water quality change.

Although the interpretation and calculation of most independent variables requires little explanation, there are some variables for which additional detail is warranted. These include variables characterizing surface water quality and its measurement. To allow the effect of water quality changes on WTP to vary systematically as a function of the primary affected species group (e.g., fish or shellfish), we included water quality in the model as a set of interactions with binary (or dummy) variables that characterized the primary species group affected, as noted in the original studies. Binary or dummy variables such as these are used in regression analysis when a value is going to be either "on" or "off" as opposed to usual continuous variables, which can take on many different values. These interaction variables distinguished the effects of water quality change for fish (*WQ_fish*), shellfish (*WQ_shell*), multiple species (*WQ_many*), and nonspecified species (*WQ_non*) (Table 7.2).

The RFF Water Quality Ladder from Mitchell and Carson (1981) is an extremely popular tool. Economists, for years, have relied on it as a means to explain to people, especially in stated choice surveys, how water quality changes.

Further explanation is also warranted for methods used to reconcile water quality measures across different studies. Many (26) observations in the meta-data characterized quality changes using variants of the Resources for the Future (RFF) water quality ladder (Mitchell and Carson 1989, p. 342). Additional details of the ladder were provided by McClelland (1974) and Vaughan (1986). This scale is linked to specific pollutant levels, which in turn are linked to the presence of aquatic species and suitability for particular recreational uses. Other observations in the meta-data, however, relied on ordinal rankings—often paired with verbal descriptions—to measure water quality. To reconcile measurements of water quality change (a prerequisite for this meta-analysis), we mapped all water quality measures to the RFF water quality ladder.

The water quality ladder allows the use of objective water quality parameters (e.g., dissolved oxygen [DO] concentrations) to characterize ecosystem services or uses provided by a given water body. Table 7.3 shows five water quality parameters, including fecal coliform (FC), DO, maximum 5-day biological oxygen demand (BOD-5), turbidity, and pH and their "minimally acceptable concentration levels" for five potential uses of a water body. Each of these water quality levels was then mapped to a 0–10 scale ladder using the weighted combinations of measured parameters. Parameter weights are provided in Table 7.3.

TABLE 7.3
Water Quality Ladder Values

Water Quality Classification		Water Quality Parameters [weights][a]				
Becomes Acceptable for	Water Quality Ladder Value	Fecal Coliform [0.242], No. Organisms/100 mL	Dissolved Oxygen [0.274], mg/L[b]	Biological Oxygen Demand-5 [0.161], mg/L	Turbidity [0.129], JTU	pH [0.194]
Drinking without treatment	9.5	0	7.0 (90)	0	5	7.25
Swimming	7.0	200	6.5 (83)	1.5	10	7.25
Game fishing	5.0	1,000	5.0 (64)	3.0	50	7.25
Rough fishing	4.5	1,000	4.0 (51)	3.0	50	7.25
Boating	2.5	2,000	3.5 (45)	4.0	100	4.25

Source: Vaughan (1986).

[a] Weights sum to 1.0 across all parameters.

[b] Percent saturation at 85°F in parentheses.

The equation underlying the scale ladder is $WQI = \prod_{i=1}^{5} q_i^{w_i}$, where WQI is the multiplicative water quality index, q_i is the quality of the ith parameter, and w_i is the weight of the ith parameter.

In most cases, the descriptions of water quality (present in the studies that did not apply the water quality ladder) rendered mapping of water quality measures to the RFF ladder straightforward. For example, studies often defined baseline and subsequent water quality in terms of suitability for recreational activities (e.g., boating, fishing, swimming) or corresponding qualitative water quality measures (e.g., poor, fair, good)—features corresponding to the RFF ladder. For studies in which such information was not provided, we used descriptive information available from studies (e.g., amount/indication of the presence of specific pollutants, historical decline of the quality of the resource) to approximate the baseline level of water quality and the magnitude of the change, based on information in Table 7.3. However, to account for potential systematic biases involved in mapping those studies that were not based on the RFF water quality ladder, we defined the binary (or dummy) variable *wq_ladder*.

This variable shows the studies in which the RFF water quality ladder measurements were an original component of the survey instrument.

THE EMPIRICAL MODEL AND RESULTS

The goal of the econometric model is to estimate model coefficients that predict the effect of each independent variable on a dependent variable representing some appropriate measure of nonmarket value. Equation (7.1) shows that the dependent variable in the present case is the natural log of stated household WTP for water quality improvements in aquatic habitat. All independent variables are linear, resulting in a semilog functional form common in meta-analysis (Johnston et al. 2005). Additional technical details of the econometric model are provided in the Appendix and in Johnston et al. Meta-analysis results are shown in Table 7.4.

Coefficient estimates in Table 7.4 reveal numerous statistically significant and intuitive patterns that influenced WTP for water quality improvements in aquatic habitats. For example, results indicate that WTP was systematically influenced by scope in various dimensions (i.e., the magnitude of resource changes), the type of habitat under consideration, the type of population sample (i.e., user vs. nonuser), and other attributes of the resources and regions in question. In general, the statistical fit of the estimated equation was good; model results suggest a considerable systematic component of WTP variation. More technical discussion of model fit, performance, and results is found in the Appendix.

SPECIFYING AND APPLYING THE BENEFIT FUNCTION

This section illustrates the development and use of a benefits transfer function based on meta-analysis results in Table 7.4. Such functions forecast WTP (or other measures of welfare) based on researcher-assigned values for model variables, chosen to represent a specific resource change and policy context. More specifically, in the present case, meta-analysis results imply a simple benefit function of the following general form:

$$\ln(WTP) = intercept + \Sigma(coefficient_i)(assigned\ variable\ value_i). \qquad (7.1)$$

Here, $\ln(WTP)$ is the dependent variable in the meta-analysis—the natural log of WTP for water quality improvements. Table 7.4 provides the estimated equation *intercept* (6.00), variable coefficients (*coefficient_i*), and the corresponding independent variable names. For example, the coefficient −0.11 corresponds to the independent variable *year_indx*, while the coefficient −1.64 corresponds to the variable *voluntary*.

As implied by Equation (7.1), once one has obtained coefficient estimates from the meta-analysis, the analyst must then assign *variable values* (i.e., choose variable levels) for each model variable. In benefits transfer applications, values for variables characterizing the resource and policy context are usually determined by the characteristics of the natural resource, site, and policy for which values are desired. Values for methodological attributes (i.e., variables characterizing the study methodology

TABLE 7.4
Meta-Analysis Results: WTP for Aquatic Habitat Improvements

Variable	Meta-Analysis Results Semilog Model Coefficient Estimate (SE)
intercept	6.0043***
	(0.6078)
year_indx	−0.1058***
	(0.0185)
discrete_ch	0.3713
	(0.3306)
voluntary	−1.6422***
	(0.2255)
interview	1.3030***
	(0.1700)
mail	0.5627***
	(0.1753)
lump_sum	0.6180***
	(0.1710)
nonparam	−0.4650**
	(0.1756)
wq_ladder	−0.3617*
	(0.1795)
protest_bids	0.9390***
	(0.1325)
outlier_bids	−0.8814***
	(0.1103)
median_WTP	0.2193
	(0.1625)
hi_response	−0.8020***
	(0.1190)
income	3.83E-07
	(4.88E-06)
nonusers	−0.5019***
	(0.1176)
single_river	−0.3236*
	(0.1791)
single_lake	0.2950
	(0.2621)
multiple_river	−1.6155***
	(0.2951)
salt_pond	0.7613**
	(0.3366)
num_rivers_ponds	0.0791***
	(0.0094)

(*continued*)

TABLE 7.4 (continued)

Variable	Meta-Analysis Results Semilog Model Coefficient Estimate (SE)
regional_fresh	−0.0069
	(0.1642)
southeast	1.1396***
	(0.2174)
pacif_mount	−0.3080**
	(0.1298)
plains	−0.7958**
	(0.2831)
mult_reg	0.6074**
	(0.2490)
WQ_fish	0.2095**
	(0.0809)
WQ_shell	0.2610**
	(0.0984)
WQ_many	0.2400**
	(0.0977)
WQ_non	0.4808**
	(0.1947)
nonfish_uses	−0.1541
	(0.1225)
fishplus	0.7964***
	(0.1719)
baseline	−0.1240***
	(0.0407)
−2 log likelihood	
Full model	65.8
Intercept and random effects only	167.6
−2 log likelihood χ^2	101.8***
Covariance Factors	
Study level (σ_u^2)	7.71×10^{-18}
Residual (σ_e^2)	0.1320
Observations (N)	81

SE, standard error.
*$p < 0.10$; **$p < 0.05$; ***$p < 0.01$.

used in the original source studies), in contrast, are often set at mean values from the meta-data. The treatment of methodological variables was discussed in more detail by Johnston et al. (2006a).

To illustrate an application of model results for benefits transfer using Equation (7.1), we applied the estimated benefit function to forecast WTP associated with increasing levels of WQ_fish (water quality improvements that primarily benefit fish and associated uses). Table 7.5 (column B) provides a set of variable levels chosen to characterize a simple, hypothetical policy scenario. The hypothetical scenario was characterized by a two-unit improvement in WQ_fish ($WQ\_fish = 2$) that occurs in a single estuary in the northeast United States ($single\_river = single\_lake = multiple\_river = salt\_pond = num\_rivers\_ponds = regional\_fresh = southeast = pacif\_mount = plains = mult\_reg = 0$). Baseline water quality was assumed to be 5.0 on the 10-point water quality scale ($baseline = 5.0$), and the water quality improvement was not expected to have a major influence on other species ($WQ\_shell = WQ\_many = WQ\_non = 0$). Nonfishing uses were assumed to be unaffected by the policy change ($nonfish\_uses=0$), and the gain in fish populations was not expected

TABLE 7.5
Using a Meta-Analysis Benefit Function to Estimate WTP

Variable	(A) Coefficient Estimates (from Table 7.4)	(B) Selected Variable Values[a]	(C) Product (A) × (B)
intercept	6.0043	1	6.0043
Study design variables			
year_indx	−0.1058	31	−3.2798
discrete_ch	0.3713	0.35	0.1299
voluntary	−1.6422	0.07	−0.1150
interview	1.3030	0.19	0.2476
mail	0.5627	0.56	0.3151
lump_sum	0.6180	0.21	0.1298
nonparam	−0.4650	0.46	−0.2139
wq_ladder	−0.3617	0.32	−0.1157
protest_bids	0.9390	0.46	0.4319
outlier_bids	−0.8814	0.22	−0.1939
median_WTP	0.2193	0.06	0.0131
hi_response	−0.8020	0.31	−0.2486
income	3.83E−07	53840	0.0206
Policy, resource, and context variables			
nonusers	−0.5019	0.19	−0.0954
single_river	−0.3236	0	0
single_lake	0.2950	0	0
multiple_river	−1.6155	0	0
salt_pond	0.7613	0	0
num_rivers_ponds	0.0791	0	0
regional_fresh	−0.0069	0	0
southeast	1.1396	0	0
pacif_mount	−0.3080	0	0
plains	−0.7958	0	0
mult_reg	0.6074	0	0
WQ_fish	0.2095	2	0.4190
nonfish_uses	−0.1541	0	0

(*continued*)

TABLE 7.5 (continued)

Variable	(A) Coefficient Estimates (from Table 7.4)	(B) Selected Variable Values[a]	(C) Product (A) × (B)
fishplus	0.7964	0	0
baseline	−0.1240	5	−0.6200
WTP estimates			
D = Sum of column (C)			2.83
$E = \sigma_e^2$ (from Table 7.4)			0.1320
WTP = $e^{(D + E/2)}$			$18.09
WTP forecasts for other values of *WQ_fish*[b]			
WTP for *WQ_fish* = 1.0		$14.67	
WTP for *WQ_fish* = 3.0		$22.30	

[a] For variables characterizing study methodology, the mean value from the meta-data (Table 7.2) was used to conduct benefits transfer (Johnston et al. 2006a). For other variables, values were selected to match the resource and policy context for which values are desired.

[b] These forecasts hold all other variable levels constant at levels shown above.

to exceed 50% (*fishplus* = 0). The percentage of nonusers in the population was assumed to be at the mean level for the meta-data (*nonusers* = 0.19), and household income was assumed to be at the median level from the 2002 Census of Population and Housing for the Northeast United States (*income* = 53,840). Methodological variables (i.e., *discrete_ch, voluntary, interview, mail, lump_sum, nonparam, wq_ladder, protest_bids, outlier_bids, median_WTP, hi_response*) were set at mean values from the meta-data (Table 7.2), except for study year, which was set at the most recent observation from the data (i.e., *year_indx* = 31, which corresponds to the year 2001). Although the present example assumed these variable values, the analyst could choose a wide range of different values to characterize alternative policy and site characteristics.

Once variable values were selected by the analyst, all that was required to forecast WTP for expected water quality improvements was simple arithmetic calculation, guided by Equation (7.1) and illustrated in Table 7.5. Coefficient estimates for each variable, taken from meta-analysis results in Table 7.4, were entered into column A of Table 7.5. Variable levels chosen were entered into column B. Column C shows the arithmetic product of columns A and B for each model variable. The sum of these products for the illustrated policy example is 2.83. This value is equivalent to the quantity (*intercept* + Σ(*coefficient$_i$*)(*assigned variable value$_i$*)) in Equation (7.1) and is given the label D in Table 7.5. This value, D = 2.83, represents the predicted *natural log of* WTP for the illustrated policy scenario, as indicated by Equation (7.1).

The final step uses a standard formula to transform this predicted natural log into the desired WTP estimate. This formula is given by

$$\text{WTP} = exp(D + \sigma_e^2/2), \tag{7.2}$$

where $exp(\cdot)$ is the exponential operator, D is defined previously, and σ_e^2 is the model error variance (0.1320) taken from Table 7.4. Applying this formula generates WTP = $18.09, which represents per-household WTP for a two-unit increase in *WQ_fish*, tailored to the specific policy context characterized (Table 7.5). This represents an estimate that could be transferred to approximate per household nonmarket value for the illustrated policy change in the absence of original study results. Multiplying this estimate by the number of households in the affected region would provide an estimate of total WTP for the policy change.

As dollar values in all source studies for the meta-analysis were adjusted to 2002 dollars prior to model estimation, the meta-analysis provides benefit estimates in 2002 dollars. Results may be adjusted to other base years by using an appropriate consumer price index (CPI) as outlined in the text box in Chapter 5.

The illustrated benefit function also allows one to conduct sensitivity analyses to assess the impact of different assumed policy outcomes or variable-level assignments on WTP. For example, assuming the same policy context, the estimated benefit function predicts that per household WTP for a one-unit increase *WQ_fish* would be equal to $14.67, while WTP for a three-unit increase would be $22.30 (Table 7.5). Other changes in model variables would generate similar changes in predicted WTP. This illustrates how a meta-analysis benefit function may be used to forecast WTP for a wide variety of different policy contexts and resource changes.

CHALLENGES AND CONCERNS

While there are many potential advantages of meta-analysis benefit functions, there are a variety of issues that must be addressed if one seeks to use such tools for applied benefits transfer. Many of these issues may not be appropriately resolved based solely on empirical considerations and involve such features as the assignment of levels for study design (or methodological) variables, methods used to reconcile environmental quality measures, the magnitude of potential transfer error, and the definition of affected populations. Such issues remain relevant, even when underlying meta-analyses have desirable statistical properties.

SENSITIVITY OF FUNCTION-BASED BENEFITS TRANSFER TO STUDY METHODOLOGY

Within a benefits transfer context, assigned values for resource and policy variables are typically determined by characteristics of the transfer site and policy context for which WTP estimates are desired, as described. These characteristics, however, do not tell us the appropriate treatment of study design variables. In practice, values for these variables are often determined based solely on analyst judgment, with little guidance from the literature (Johnston et al. 2005). For example, researchers sometimes choose study design variable levels based on guidance regarding the general appropriateness of particular research methods for welfare evaluation or specify these variables at mean values from the meta-data (Johnston et al. 2006a). The latter approach was illustrated. Although the appropriateness of such approaches may vary across data sets and policy contexts, the practical implications for benefits transfer can be substantial as the magnitude of WTP transfer can vary considerably according to these choices. Examples of such patterns were demonstrated by Johnston et al. (2005, 2006a,b). As a result, analysts conducting function-based benefits transfer should be cognizant of the potential importance of study design variables for resulting benefit estimates.

RECONCILIATION OF ENVIRONMENTAL QUALITY MEASURES

Similar sensitivity of WTP may be shown to choices made regarding the reconciliation of environmental quality measures across studies. For example, differences in methods used to represent water quality change in original studies are captured by wq_ladder; $wq\_ladder = 1$ implies that the original study used a variant of the RFF water quality ladder (Mitchell and Carson 1989, p. 342); $wq\_ladder = 0$ implies that the original study did not use the ladder, and that water quality measures from the study were manually mapped to the ladder prior to conducting the meta-analysis as discussed. This reconciliation was required so that water quality measures could be compared across studies. Meta-analysis results, however, indicated that lower WTP was associated with the use of the water quality ladder in the original survey.

For example, the illustration shows that WTP for a two-unit increase in WQ_fish is equal to $18.09, given the policy context characterized by Table 7.5. This estimate assumes that $wq\_ladder = 0.32$, which is the mean value of this variable from the meta-data. However, one might also argue that $wq\_ladder = 1$ is a more appropriate assignment as this reflects values from studies in which the water quality ladder was a native part of the research effort, and there may have been systematic biases involved in mapping results from nonladder studies onto the water quality ladder. If one makes this assumption, then WTP for the same resource change decreases to $14.14—a 21.8% decrease from the original estimate of $18.09.

The sensitivity of WTP to variables such as wq_ladder—which account for mechanisms used to reconcile quality measures—has obvious implications for benefits transfer. However, the appropriate response to such sensitivity is unclear and may depend on the suspected rationale for the statistical significance of such variables. For example, if the significance of wq_ladder were related to a true reduction in WTP associated with the use of the RFF water quality ladder in survey instruments, then an appropriate action might be to specify the variable at its mean value unless one has a preconceived reason to believe that lower WTP estimates associated with ladder-using studies are more appropriate. In contrast, if one suspects that the significance of this variable is due to systematic biases involved in mapping water quality measurements from those studies that are not based on the RFF water quality ladder, then a more appropriate action might be to set $wq\_ladder = 1$ to offset this suspected bias.

In the present case, the rationale for the statistical significance is ambiguous, leading to uncertainty regarding the appropriate treatment of this variable in a benefits transfer context. As shown, ambiguity in the treatment of such effects may have substantial implications for the outcome of a benefits transfer exercise.

MAGNITUDE OF TRANSFER ERROR

Transfer error may be defined as the error that occurs when benefit estimates from a study site (or combination of sites) are used or adapted to forecast benefits at a policy site; it is the difference between the transferred and actual, generally unknown, value (Rosenberger and Stanley 2006). The magnitude of such errors is critical to the validity and accuracy of benefits transfer. Past research showed that transfer errors may be reduced by transferring benefit functions rather than point estimates of value (Rosenberger and Stanley 2006), and benefit functions estimated with meta-analysis usually do pretty well, but there are some cases when other methods are better. This may be particularly true when meta-analyses are based on small samples of research studies. As a result, practitioners should not assume that meta-analysis will *always* represent the preferred means of conducting function-based benefits transfer. The appropriateness of meta-analysis for applied benefits transfer, and the extent of error that is to be expected, will depend on elements such as linkages between the estimated benefit function and established economic theory, the quality and characteristics of the underlying data, and the performance of the statistical equation (Bergstrom and Taylor 2006).

Defining Affected Populations

Meta-analysis results shown allowed estimation of per household WTP for specified water quality changes. However, evaluating the *total value* of water quality improvements also requires a definition of the size of the population expected to hold these values. That is, total value is equivalent to average per household value multiplied by the number of affected households. In some cases, the affected population may include both users and nonusers of the affected resources. There has been some research addressing the size of populations that hold values for particular types of resource changes (e.g., Pate and Loomis 1997; Schulze et al. 1995). Despite this research, most decisions regarding the size of affected populations—required for benefits transfer using meta-analysis—are made on a case-by-case basis, often based on site-specific or ad hoc criteria. For example, benefits transfer for a statewide water quality improvement program might assume an affected population of all state households. Alternatively, one might conduct sensitivity analysis to assess the impact of assumptions regarding affected populations on the size of total transferred benefits; such analyses are often the most appropriate solution when affected populations are uncertain. In most cases, however, some degree of researcher judgment is required to determine the population for which per household values should be aggregated; this adds another layer of complexity to many benefits transfer applications.

CONCLUSION

Although the appropriate estimation of meta-analysis models requires a fair degree of expertise, benefits transfer based on already-estimated meta-analyses may be conducted by those with less training. It was the goal of this chapter to illustrate simple methods by which practitioners may use appropriate meta-analysis results to conduct benefits transfer when benefit estimates are required but original valuation studies are unavailable. The chapter provided a basic understanding of meta-analysis benefits transfer suitable for those without extensive training in applied valuation and to highlight some of the primary challenges.

We emphasize that this chapter did not provide complete coverage of the many challenges involved in the appropriate use of benefits transfer for policy guidance. There is a substantial research literature addressing various aspects of benefits transfer, with particular stress on conditions under which transfer estimates are expected to provide sufficient approximations of underlying, true benefits (Wilson and Hoehn 2006). Notwithstanding this extensive and growing literature, "few benefits transfer practitioners seem fully satisfied with the state of [benefits transfer] science and continue to strive for agreement on best practice standards" (Wilson and Hoehn 2006, p. 336).

Recognizing the potential pitfalls that can face benefits transfer practitioners, the focus on simple application shown in this chapter is in no way meant to downplay the need for technical expertise in the estimation of meta-analysis models and the use of these models for benefits transfer. The validity of any benefits transfer based on a meta-analysis depends critically on the empirical quality of the underlying meta-analysis and correspondence to appropriate economic theory (Bergstrom and Taylor 2006). As noted, there is also a variety of challenges faced in the estimation and interpretation of meta-models for applied use. These challenges involve the sensitivity of transfer estimates to such factors as the treatment of study design variables and methods used to reconcile environmental quality measures and remain salient even when the statistical performance of meta-models is exemplary. Those seeking to apply meta-analysis for benefits transfer should be aware of such issues and their potential implications for the appropriateness, validity, and accuracy of applied transfers. As a result, it is generally advisable to consult with experts when seeking to conduct benefits transfers for policy guidance. These concerns notwithstanding, meta-analysis can represent a significant addition to the toolbox available for applied benefits transfer and can represent a viable means to approximate resource values in a variety of policy settings.

APPENDIX: DETAILS OF THE MODEL AND ESTIMATED RESULTS

Links to Underlying Utility

The illustrated meta-analysis is conceptualized as a "weak structural utility theoretic" approach in which the "connection between explanatory variables and an underlying utility function are explicitly specified, but only as approximations" (Bergstrom and

Taylor 2006, p. 352). To illustrate these links, we specify conditional indirect utility for individual j as a general function:

$$V_j = V_j(Q_j, A_j, M_j, P_j, S_j, D_j)$$ (7.A1)

Here, Q_j is water quality in a particular water body, A_j is a vector of other attributes of the water body, M_j is household income, P_j is a vector of exogenous prices, S_j is a vector of attributes characterizing other (e.g., substitute) resources and other exogenous conditions, and D_j is a vector of nonincome household attributes.

Based on this simple specification, compensating surplus or WTP for an increase in Q_j, assuming no other exogenous changes, is characterized by

$$V_j(Q_j^0, A_j, M_j, P_j, S_j, D_j) = V_j(Q_j^1, A_j, M_j - WTP_j, P_j, S_j, D_j).$$ (7.A2)

where Q_j^0 is original water quality, and Q_j^1 is subsequent, improved quality. Solving for WTP_j results in the general function

$$WTP_j = f(Q_j^0, Q_j^1, A_j, M_j, P_j, S_j, D_j).$$ (7.A3)

If we further assume that WTP for environmental improvements may be modeled as separable from market prices, Equation (7.A3) simplifies to

$$WTP_j = f(Q_j^0, Q_j^1, A_j, M_j, S_j, D_j).$$ (7.A4)

This is the general form of the valuation or benefit function that is used as guidance in specifying the empirical benefit function illustrated in Table 7.4. More specifically, independent variables in the benefit function are chosen to represent key attributes in the theoretical valuation function (7.A4). The statistical model is considered an empirical approximation of the theoretical equation (7.A4), with the specific econometric functional form chosen on empirical grounds (e.g., statistical fit).

ECONOMETRIC MODEL AND RESULTS

Complete technical details of the statistical model are found in Johnston et al. (2005), including a comparison of different model specifications. The semilog functional form was chosen based on its statistical performance and ability to capture curvature in the valuation function and because it allows independent variables to influence WTP in a multiplicative rather than additive manner. We applied a random effects model to the meta-data to address potential correlation among observations gathered from single studies. We also applied robust variance estimation; this "approach treats each study as the equivalent of a sample cluster with the potential for heteroskedasticity ... across clusters" (Smith and Osborne 1996, p. 293). All observations in the meta-data were given equal weight in the analysis.

Results are provided in Table 7.4, as highlighted in the main text. Likelihood ratio tests (Table 7.4) showed that model variables were jointly significant at $p < .01$.

The majority of independent variables were statistically significant at $p < .10$, with most statistically significant at $p < .01$. Considering these factors, the statistical performance of the model compared favorably to prior meta-analyses in the valuation literature. While the model provides evidence of systematic WTP variation associated with resource, context, and study attributes, random effects associated with systematic study-level variance (σ_u^2) were not statistically significant. This finding is similar to those of Bateman and Jones (2003) and Johnston et al. (2003) and suggests that once one accounts for variation in observable resource, context, and study attributes, no additional systematic variation in WTP may be ascribed to study-level effects. This finding suggests that systematic variation in WTP is not driven by unobservable attributes unique to particular studies or sets of study authors. The following sections highlight some of the primary systematic patterns identified by the meta-analysis and their relationships to prior expectations and findings in the literature.

SYSTEMATIC COMPONENTS OF WTP: RESOURCE ATTRIBUTES

The variables *WQ_fish*, *WQ_shell*, *WQ_many*, and *WQ_non* indicated the effects of water quality improvements associated with gains in fish, shellfish, multiple species, and unspecified habitat, respectively (Table 7.2). All signs were as expected. The associated coefficients were positive and statistically significant ($p < .02$ or better), indicating that higher WTP was associated with larger gains in water quality as measured on the RFF ladder (Table 7.4). This is a noteworthy result as it indicates that WTP—compared systematically across studies—is sensitive to the scope of water quality improvements (cf. Smith and Osborne 1996; Johnston et al. 2003).

Results also suggest that WTP for water quality improvement declined as baseline water quality increased. The variable *baseline* represents the baseline water quality from which water quality change would occur. The associated parameter estimate was significant ($p < .01$) and of the expected negative sign, revealing diminishing returns to scale for water quality improvements. This finding suggests that WTP across studies is systematically sensitive to scope not only at a broad level (i.e., larger water quality improvements generate larger WTP), but also at a more subtle, if no less important, level associated with diminishing marginal returns to scale. Finally, the variable *fishplus* identified those studies for which the associated survey identified particularly large gains in fish populations or harvest rates (>50%). The positive and statistically significant result ($p < .01$) indicated that particularly large gains in fish populations or harvests were associated with statistically significant increases in total WTP.

SYSTEMATIC COMPONENTS OF WTP: GEOGRAPHICAL AND WATER BODY TYPE ATTRIBUTES

Ten binary variables characterized geographic region and scale and water body type; eight were statistically significant at $p < .10$. The default category from which these variables allow systematic variations in WTP is an estuarine water body in the northeast United States. Compared to this baseline, lower WTP was associated with rivers (*single_river*, *multiple_river*), while higher WTP was associated with water quality gains in salt ponds (*salt_pond*). *Single_lake* and *regional_fresh* both had positive values, but neither was statistically significant.

Results further suggested that WTP was sensitive to the number of water bodies under consideration. Of the water body categories distinguished, both rivers and salt ponds included variation in numbers of affected water bodies explicitly described by the survey. This variation was captured by the variable *num_riv_pond* (Table 7.2). The associated parameter estimate was statistically significant ($p < .01$) and indicated that WTP increased with the number of water bodies considered (Table 7.4). This result, combined with the statistical significance of the water quality change variables noted, suggests that WTP values in the meta-data were sensitive to scope—in terms of both the number of water bodies and the magnitude of quality change. Such multidimensional scope sensitivity extended findings such as those of Smith and Osborne (1996), which addressed sensitivity to scope in more limited dimensions.

Finally, the regional indicator variables *southeast*, *pacif_mount*, *plains*, and *mult_reg* were statistically significant at $p < .05$ (most at $p < .01$), suggesting that there are significant differences among WTP estimates from surveys in different geographical regions of the United States. While such effects may be related to systematic differences in preferences or resource characteristics across regions, they may also be related to otherwise unexplained characteristics of authors, methodology, or other factors that may be correlated with geographical region.

Systematic Components of WTP: Population Attributes

WTP studies often differ with regard to the presence and type of demographic and other variables that characterize sampled populations. Given the wide disparity in the treatment of such factors, meta-analyses in the valuation literature typically included relatively few variables that characterized sampled populations (e.g., Poe et al. 2001; Smith and Osborne 1996). Here, only two variables, *nonusers* and *income*, were used to characterize surveyed populations. The variable *nonusers* was of particular relevance. The negative and significant ($p < .01$) parameter estimate indicated that surveys of nonusers only—where nonusers by definition had only nonuse values for the resource improvements in question (Freeman 2003, p. 142)—generated lower WTP values than surveys that included users, who may have had both use and nonuse values.

Systematic Components of WTP: Study Attributes

A variety of study and methodology effects may be shown to influence WTP for water quality improvements. While not surprising, this does indicate that methodological approach influences WTP, as indicated by prior meta-analyses (e.g., Johnston et al. 2003; Brouwer 2002; Rosenberger and Loomis 2000a; Smith and Osborne 1996).

Of 12 variables characterizing study and methodological effects, 10 had statistically significant effects on WTP. Among these was the year in which a study was conducted (*year_indx*), with later studies associated with lower WTP. This was an expected result as the focus of stated preference survey design over time has often been on the reduction of survey biases that would otherwise result in an overstatement of WTP (Arrow et al. 1993).

Model results revealed that voluntary (*voluntary*) payment vehicles (i.e., surveys that described hypothetical payments as voluntary) were associated with reduced WTP estimates. This result counters common intuition that voluntary payment vehicles may be associated with overstatements of true WTP. The reason for this finding is unknown but may indicate an unwillingness among respondents to proffer large voluntary payments given the fear that others will get a free ride. Reduced WTP estimates were also associated with studies applying nonparametric methods (*nonparam*). Survey response format (e.g., *discrete_ch*) did not have a statistically significant effect in the model.

Smaller WTP estimates were associated with studies that eliminated or trimmed outlier bids when estimating WTP (*outlier_bids*). Conversely, increased WTP estimates were associated with studies that sought to eliminate protest bids (*protest_bids*), suggesting a preponderance of zero protest bids. Especially when eliciting values that relate to ecological resources, such as fish species, such bids may be provided by respondents that have preference structures at variance with consumer choice axioms; they may be essentially unwilling to equate an ecological change with *any* dollar amount (Spash 2000).

Studies with high response rates (*hi_response*) were associated with lower WTP estimates, an expected result associated with limiting avidity bias. In addition, lower WTP was associated with the use of the RFF water quality ladder in the original survey (*wq_ladder*). As is the case with a variety of study design variables, there was no necessary expectation with respect to the direction of this effect. Survey format variables also had an effect on WTP, as might be expected. *Interview* and *mail* both had positive and statistically significant coefficients compared to the default of telephone surveys.

WTP values for the majority of studies included in the analysis consisted of annual payments over an indefinite duration. However, a small number of studies estimated WTP for payments over a short horizon—typically 3 to 5 years. The variable *lump_sum* identified studies in which payments were to occur on something other than an indefinite annual basis (Table 7.2). The positive and statistically significant parameter for *lump_sum* indicated sensitivity to the payment schedule (Stevens et al. 1997). Studies that asked respondents to report an annual payment (as opposed to a shorter *lump_sum* payment) had lower nominal WTP estimates.

ACKNOWLEDGMENT

This research was partially funded under USEPA contract 68-C-99-239. Opinions belong solely to the authors and do not necessarily reflect the views or policies of USEPA or imply endorsement by the funding agency.

NOTES

1. However, others are more cautious regarding the use of this method for certain applications (e.g., Poe et al. 2001).
2. There are some exceptions to this rule, as discussed by Heberlein et al. (2005).
3. The Appendix illustrates the relationship between meta-analysis variables and the presumed indirect utility function, following Bergstrom and Taylor (2006).

REFERENCES

Aiken, R.A. 1985. Public Benefits of Environmental Protection in Colorado. Master's thesis, Colorado State University, Fort Collins, CO.

Anderson, G.D., and S.F. Edwards. 1986. Protecting Rhode Island's coastal salt ponds: An economic assessment of downzoning. *Coastal Zone Management* 14(1/2): 67–91.

Arrow, K., R. Solow, E. Leamer, P. Portney, R. Rander, and H. Schuman. 1993. Report of the NOAA Panel on Contingent Valuation. *Federal Register* 58(10): 4602–4614.

Azevedo, C., J.A. Herriges, and C.L. Kling. 2001. *Valuing Preservation and Improvements of Water Quality in Clear Lake.* Center for Agricultural and Rural Development (CARD), Iowa State University, Staff Report 01-SR 94, Iowa State University, Ames, IA.

Bateman, I.J., and A.P. Jones. 2003. Contrasting conventional with multi-level modeling approaches to meta-analysis: Expectation consistency in U.K. woodland recreation values. *Land Economics* 79(2): 235–258.

Bergstrom, J.C., and P. De Civita. 1999. Status of benefits transfer in the United States and Canada: A review. *Canadian Journal of Agricultural Economics* 47(1): 79–87.

Bergstrom, J.C., and L.O. Taylor. 2006. Using meta-analysis for benefits transfer: Theory and practice. *Ecological Economics* 60(2): 351–360.

Bockstael, N.E., K.E. McConnell, and I.E. Strand. 1989. Measuring the benefits of improvements in water quality: The Chesapeake Bay. *Marine Resource Economics* 6(1): 1–18.

Brouwer, R. 2002. Environmental value transfer: State of the art and future prospects, in *Comparative Environmental Economic Assessment*, P. Nijkamp, R. Florax, and P. Willis, Eds. Cheltenham, U.K.: Elgar, pp. 90–116.

Button, K. 2002. An evaluation of the potential of meta-analysis, in *Comparative Environmental Economic Assessment*, P. Nijkamp, R. Florax, and P. Willis, Eds. Cheltenham, U.K.: Elgar, pp. 74–89.

Cameron, T.A., and D.D. Huppert. 1989. OLS versus ML estimation of non-market resource values with payment card interval data. *Journal of Environmental Economics and Management* 17(3): 230–246.

Carson, R.T., W.M. Hanemann, R.J. Kopp, J.A. Krosnick, R.C. Mitchell, S. Presser, P.A. Ruud, and V.K. Smith. 1994. *Prospective Interim Lost Use Value due to DDT and PCB Contamination in the Southern California Bight.* Vol. 2. Report to the National Oceanic and Atmospheric Administration. La Jolla, CA: Natural Resources Damage Assessment.

Clonts, H.A., and J.W. Malone. 1990. Preservation attitudes and consumer surplus in free flowing rivers, in *Social Science and Natural Resource Recreation Management*, J. Vining, Ed. Boulder, CO: Westview Press, pp. 301–317.

Croke, K., R.G. Fabian, and G. Brenniman. 1987. Estimating the value of improved water quality in an urban river system. *Journal of Environmental Systems* 16(1): 13–24.

Cronin, F.J. 1982. *Valuing Nonmarket Goods Through Contingent Markets.* PNL-4255. Richland, WA: Pacific Northwest Laboratory.

Desvousges, W.H., V.K. Smith, and M.P. McGivney. 1983. *A Comparison of Alternative Approaches for Estimating Recreation and Related Benefits of Water Quality Improvements.* Washington, DC: U.S. Environmental Protection Agency, Economic Analysis Division.

De Zoysa, A.D.N. 1995. A Benefit Evaluation of Programs to Enhance Groundwater Quality, Surface Water Quality and Wetland Habitat in Northwest Ohio. Dissertation, Ohio State University, Columbus, OH.

Farber, S., and B. Griner. 2000. Valuing watershed quality improvements using conjoint analysis. *Ecological Economics* 34(1): 63–76.

Freeman, A.M., III. 2003. *The Measurement of Environmental and Resource Values: Theory and Methods.* Washington, DC: Resources for the Future.

Glass, G.V. 1976. Primary, secondary, and meta-analysis of research. *Educational Researcher* 5(10): 3–8.

Hayes, K.M., T.J. Tyrrell, and G. Anderson. 1992. Estimating the benefits of water quality improvements in the Upper Narragansett Bay. *Marine Resource Economics* 7: 75–85.

Heberlein, T.A., M.A. Wilson, R.C. Bishop, and N.C. Schaeffer. 2005. Rethinking the Scope Test as a criterion for validity in contingent valuation. *Journal of Environmental Economics and Management* 50(1): 1–22.

Herriges, J.A., and J.F. Shogren. 1996. Starting point bias in dichotomous choice valuation with follow-up questioning. *Journal of Environmental Economics and Management* 30(1): 112–131.

Huang, J.C., T.C. Haab, and J.C. Whitehead. 1997. Willingness to pay for quality improvements: Should revealed and stated preference data be combined? *Journal of Environmental Economics and Management* 34(3): 240–255.

Johnston, R.J., E.Y. Besedin, R. Iovanna, C. Miller, R. Wardwell, and M. Ranson. 2005. Systematic variation in willingness to pay for aquatic resource improvements and implications for benefits transfer: A meta-analysis. *Canadian Journal of Agricultural Economics* 53(2–3): 221–248.

Johnston, R.J., E.Y. Besedin, and M.H. Ranson. 2006a. Characterizing the effects of valuation methodology in function-based benefits transfer. *Ecological Economics* 60(2): 407–419.

Johnston, R.J., E.Y. Besedin, and R.F. Wardwell. 2003. Modeling relationships between use and nonuse values for surface water quality: A meta-analysis. *Water Resources Research* 39(12): 1363–1372.

Johnston, R.J., M.H. Ranson, E.Y. Besedin, and E.C. Helm. 2006b. What determines willingness to pay per fish? A meta-analysis of recreational fishing values. *Marine Resource Economics* 21(1): 1–32.

Kaoru, Y. 1993. Differentiating use and nonuse values for coastal pond water quality improvements. *Environmental and Resource Economics* 3: 487–494.

Lant, C.L., and R.S. Roberts. 1990. Greenbelts in the cornbelt: Riparian wetlands, intrinsic values, and market failure. *Environment and Planning A* 22(10): 1375–1388.

Loomis, J.B. 1996. How large is the extent of the market for public goods: Evidence from a nationwide contingent valuation survey. *Applied Economics* 28(7): 779–782.

Lyke, A.J. 1993. Discrete Choice Models to Value Changes in Environmental Quality: A Great Lakes Case Study. Dissertation, University of Wisconsin–Madison, Madison, WI.

Magat, W.A., J. Huber, K.W. Viscusi, and J. Bell. 2000. An iterative choice approach to valuing clean lakes, rivers, and streams. *Journal of Risk and Uncertainty* 21(1): 7–43.

Matthews, L.G., F.R. Homans, and K.W. Easter. 1999. *Reducing Phosphorous Pollution in the Minnesota River: How Much is it Worth?* Staff Paper. Department of Applied Economics, University of Minnesota, St. Paul, MN.

McClelland, N.I. 1974. *Water Quality Index Application in the Kansas River Basin.* Report to USEPA Region VII. Kansas City, MO.

Mitchell, R.C., and R.T. Carson. 1981. *An Experiment in Determining Willingness to Pay for National Water Quality Improvements.* Preliminary Draft of a Report to the U.S. Environmental Protection Agency. Washington, DC: Resources for the Future.

Mitchell, R.C., and R.T. Carson. 1989. *Using Surveys to Value Public Goods: The Contingent Valuation Method.* Washington, DC: Resources for the Future.

Olsen, D., J. Richards, and R.D. Scott. 1991. Existence and sport values for doubling the size of Columbia River Basin salmon and steelhead runs. *Rivers* 2(1): 44–56.

Pate, J., and J. Loomis. 1997. The effect of distance on willingness to pay values: A case study of wetlands and salmon in California. *Ecological Economics* 20(3): 199–207.

Poe, G.L., K.J. Boyle, and J.C. Bergstrom. 2001. A preliminary meta analysis of contingent values for ground water quality revisited, in *The Economic Value of Water Quality,* J.C. Bergstrom, K.J. Boyle, and G.L. Poe, Eds. Northampton, MA: Elgar, pp. 137–162.

Roberts, L.A., and J.A. Leitch. 1997. *Economic Valuation of Some Wetland Outputs of Mud Lake.* Agricultural Economics Report No. 381. Department of Agricultural Economics, North Dakota Agricultural Experiment Station, North Dakota State University. Fargo, ND.

Rosenberger, R.S., and J.B. Loomis. 2000a. Panel stratification in meta-analysis of economic studies: An investigation of its effects in the recreation valuation literature. *Journal of Agricultural and Applied Economics* 32(3): 459–470.

Rosenberger, R.S., and J.B. Loomis. 2000b. Using meta-analysis for benefits transfer: In-sample convergent validity tests of an outdoor recreation database. *Water Resources Research* 36(4): 1097–1107.

Rosenberger, R.S., and J.B. Loomis. 2003. Benefits transfer, in *A Primer on Non-Market Valuation,* P.A. Champ, K.J. Boyle, and T.C. Brown, Eds. pp. 445–482. Dordrecht, The Netherlands: Kluwer Academic.

Rosenberger, R.S., and T.D. Stanley. 2006. Measurement, generalization and publication: Sources of error in benefits transfers and their management. *Ecological Economics* 60(2): 372–378.

Rowe, R.D., W.D. Schulze, B. Hurd, and D. Orr. 1985. *Economic Assessment of Damage Related to the Eagle Mine Facility.* Boulder, CO: Energy and Resource Consultants.

Sanders, L.B., R.G. Walsh, and J.B. Loomis. 1990. Toward empirical estimation of the total value of protecting rivers. *Water Resources Research* 26(7): 1345–1357.

Schulze, W.D., R.D. Rowe, W.S. Breffle, R.R. Boyce, and G.H. McClelland. 1995. *Contingent Valuation of Natural Resource Damages Due to Injuries to the Upper Clark Fork River Basin,* State of Montana, Natural Resource Damage Litigation Program. Boulder, CO: RCG/Hagler Bailly.

Smith, V.K., G. Van Houtven, and S.K. Pattanayak. 2002. Benefits transfer via preference calibration: "Prudential Algebra" for policy. *Land Economics* 78(1): 132–152.

Smith, V.K., and L. Osborne. 1996. Do contingent valuation estimates pass the Scope Test? A meta analysis. *Journal of Environmental Economics and Management* 31(3): 287–301.

Spash, C.L. 2000. Ecosystems, contingent valuation and ethics: The case of wetland re-creation. *Ecological Economics* 34(2): 195–215.

Stevens, T., N. De Coteau, and C. Willis. 1997. Sensitivity of contingent valuation to alternative payment schedules. *Land Economics* 73(1): 140–148.

Stumborg, B.E., K.A. Baerenklau, and R.C. Bishop. 2001. Nonpoint source pollution and present values: A contingent valuation of Lake Mendota. *Review of Agricultural Economics.* 23(1): 120–132.

Sutherland, R.J., and R.G. Walsh. 1985. Effect of distance on the preservation value of water quality. *Land Economics* 61(3): 282–291.

U.S. Census Bureau. 2002. 2002 Census, Summary File 3: United States. U.S. Census Bureau: Washington, DC.

U.S. Environmental Protection Agency (USEPA). 2000. *Guidelines for Preparing Economic Analyses.* EPA 240-R-00-003. Washington, DC: USEPA, Office of the Administrator.

Vaughan, W.J. 1986. The RFF water quality ladder, in *The Use of Contingent Valuation Data for Benefit/Cost Analysis in Water Pollution Control, Final Report,* R.C. Mitchell and R.T. Carson, Eds. Appendix B. Washington, DC: Resources for the Future.

Welle, P.G. 1986. Potential Economic Impacts of Acid Deposition: A Contingent Valuation Study of Minnesota. Dissertation, University of Wisconsin–Madison, Madison, WI.

Wey, K.A. 1990. Social Welfare Analysis of Congestion and Water Quality of Great Salt Pond, Block Island, Rhode Island. Dissertation, University of Rhode Island, Kingston, RI.

Whitehead, J.C., and G.C. Blomquist. 1991. A link between behavior, information, and existence value. *Leisure Sciences* 13(2): 97–109.

Whitehead, J.C., G.C. Blomquist, T.J. Hoban, and W.B. Clifford. 1995. Assessing the validity and reliability of contingent values: A comparison of on-site users, off-site users, and non-users. *Journal of Environmental Economics and Management* 29(2): 238–251.

Whitehead, J.C., and P.A. Groothuis. 1992. Economic benefits of improved water quality: A case study of North Carolina's Tar-Pamlico River. *Rivers* 3: 170–178.

Whittington, D., G. Cassidy, D. Amaral, E. McClelland, H. Wang, and C. Poulos. 1994. *The Economic Value of Improving the Environmental Quality of Galveston Bay.* GBNEP-38, 6/94. Chapel Hill, NC: Department of Environmental Sciences and Engineering, University of North Carolina at Chapel Hill.

Wilson, M.A., and J.P. Hoehn. 2006. Valuing environmental goods and services using benefits transfer: The state-of-the art and science. *Ecological Economics* 60(2): 335–342.

8 Using Conceptual Models to Communicate Environmental Changes

*Matthew T. Heberling, George Van Houtven,
Stephen Beaulieu, Randall J. F. Bruins,
Evan Hansen, Anne Sergeant, and
Hale W. Thurston*

CONTENTS

INTRODUCTION

As defined in Chapter 1, economic value is based on change (e.g., in environmental quality). The change could be actual or proposed, from implementing a management alternative (e.g., planting a grass filter strip along a farm field near a stream to reduce runoff) to creating a new source of pollution. One potential tool that helps illustrate and communicate the changes is the conceptual model; it is a visual representation along with written descriptions of the relationships among sources or land uses, stressors (or pollutants), ecological entities, and their responses to the stressors. In this chapter, the typical conceptual model, as found in ecological risk assessment (e.g., see U.S. Environmental Protection Agency [USEPA] 1998), is expanded to include ecosystem services (e.g., ecosystem functions or processes that directly or indirectly affect an individual's happiness or well-being), water quality standards (WQS), restoration actions, and economic endpoints.

Conceptual models are not necessary for valuation exercises, but they are useful both to guide technical evaluation of alternatives in economic analyses and to communicate the changes to stakeholders or watershed associations. They should help stakeholders appreciate the relative merits of management and restoration alternatives, in terms of both the costs of the actions and their expected ecological benefits.

BACKGROUND

We base this chapter on previous research (USEPA 2007b) that developed methods to assist communities with decisions about changes to WQS, specifically the water quality goals or designated uses. The general approach presented here for developing and presenting the conceptual models is the same for a watershed group attempting to model specific restoration alternatives. The conceptual model may not need all of the elements included; in fact, it should be tailored with more or less details in the diagrams and text depending on the audience.

The establishment of WQS requires determining the designated uses of each regulated water body. Designated uses protect the natural integrity of the nation's waters and their uses by people and aquatic organisms (e.g., propagation of aquatic life, drinking water, recreational contact, etc.). However, the Clean Water Act (CWA) also recognized that, in some cases, states or tribes must evaluate changes to a designated use—for example, because naturally occurring, human-made, or socioeconomic factors inhibit its attainment or, conversely, because a higher use is attainable. In certain cases, decisions related to changing or attaining designated uses involve gains and losses (herein called *trade-offs*) among health, ecological, and socioeconomic considerations. States and tribes are provided limited latitude in adopting or revising designated uses and must balance these trade-offs carefully.

The following are examples of trade-offs: A significant reduction in the discharge of pollutants to a stream might restore a blue ribbon trout fishery and make the stream safe for full-contact recreation such as swimming, but it also may require a substantial increase in treatment costs. On the other hand, a modest reduction with a modest increase in treatment costs may allow the stream to support trout year round yet make the water only safe enough for incidental contact recreation such as fishing and boating.

When changes to designated uses are contemplated, states or tribes are required to conduct use attainability analyses (UAAs) or variance analyses. The purpose of these scientific assessments is to examine the factors that may be affecting the attainment of a designated use. A variance analysis, similar to a UAA, may be conducted to obtain a temporary relaxation of the WQS. In other cases, states and tribes may consider permitting a reduction of water quality in high-quality waters if the reduced quality will not affect designated uses. Under these conditions, the CWA requires formal antidegradation reviews to demonstrate that the reduction is necessary to accommodate important economic or social development in the area. Thus, the ultimate determination of water quality goals for a stream, lake, or estuary may require the evaluation of both ecological and socioeconomic objectives.

UNDERSTANDING ECOSYSTEM SERVICES

For watershed groups to understand the broader ramifications of alternative restoration options, it is often necessary to look beyond the financial impacts. Therefore, an objective of this chapter is to help watershed groups and decision makers better understand how humans interact with and derive services from the affected ecological systems and how these services are related to WQS management options and designated uses. To do this, the following sections define and describe aquatic ecosystem services and their relationship to designated uses. Even if designated uses are not important to the watershed group, the ecosystem services that do matter to them are still likely to be affected by management alternatives or changes in pollutant levels.

AQUATIC ECOSYSTEM SERVICES

Understanding the concept of ecosystem services is fundamental for evaluating how humans are supported by ecological systems and how their well-being is affected by changes in these systems (see, e.g., Daily 1997 or Millennium Ecosystem Assessment 2005). This chapter adopts the following definition provided by USEPA (2006, p. 4):

> **Ecosystem services** are outputs of ecological functions or processes that directly or indirectly contribute to social welfare or have the potential to do so in the future. Some may be bought and sold, but most are not marketed.

The concept of *aquatic ecosystem services* is particularly important for the purpose of setting and evaluating WQS and goals for watershed groups. These are the services specifically derived from surface water resources and their connected ecosystems. They are also the ecosystem services primarily affected by alternative water quality management options.

Figure 8.1 illustrates the link between aquatic ecosystems and the services derived from these systems. It provides the most basic description of the primary components and processes of a functioning aquatic ecosystem. These include the physical habitat (e.g., stream bed characteristics and the flow of water through the system), the biological components (e.g., fish populations and species diversity), and the chemical, biological, and hydrological processes that occur within the ecosystem. These components and processes directly influence and are influenced by the level of water quality (e.g., dissolved oxygen content and pH levels) in the system.

Figure 8.1 shows that the interrelated features of an aquatic ecosystem provide a wide range of ecosystem services to humans. These services typically are derived from specific human uses of surface water resources and their associated aquatic ecosystems. The uses include activities that are primarily commercial, such as commercial fishing, navigation, energy production, and agriculture (e.g., through crop irrigation). In addition, the services also include "nonmarket" activities, which are

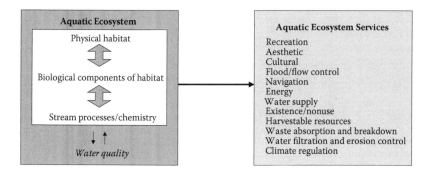

FIGURE 8.1   Aquatic ecosystems and examples of services derived from these systems.

unrelated or only indirectly related to commercial activities, such as water-based recreation, subsistence fishing, and household water use. Other services provided by aquatic ecosystems relate to or support a wide variety of human uses. For example, flood control services protect commercial and residential properties as well as water-based recreational facilities. Aesthetic services from aquatic ecosystems (e.g., through appreciation of their natural beauty) enhance recreational, residential, and many other uses of water resources.

Only one of the ecosystem service categories—existence/nonuse—is, by definition, unrelated to any specific human uses of water resources. The argument for including existence/nonuse as a distinct category of ecosystem service is that individuals can gain satisfaction and fulfillment simply from the knowledge that an ecosystem (particularly a well-functioning and healthy one) exists. For example, these services can arise because individuals

- value the ecosystem intrinsically
- value the satisfaction others get from using the resource (altruistic value)
- value preserving the resource for future generations (bequest/preservation value)
- gain satisfaction from a sense of environmental stewardship

GENERAL FRAMEWORK FOR THE EXPANDED CONCEPTUAL MODELS

Building on Figure 8.1, Figure 8.2 shows that land uses and other sources are capable of introducing stressors to aquatic ecosystems. These stressors can disrupt any number of processes in a functioning ecosystem, which can cause reductions in water quality and can impair the ecosystem's ability to provide key services. However, these same sources and land uses are also capable of providing other important goods and services to humans. For example, agricultural land uses may degrade water quality in local streams while providing valued food crops for consumers.

Figure 8.3 further extends the framework by illustrating how management options considered in a standard-setting process will typically alter the effects of land uses and sources on human well-being (e.g., restoring a riparian area or building

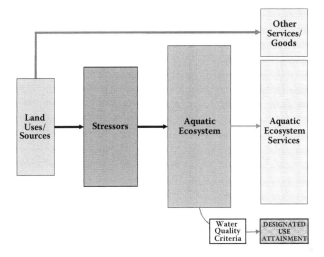

FIGURE 8.2 Effects of sources/stressors on aquatic ecosystem services, use attainment, and provision of other goods and services.

a stormwater retention pond). Because both gains and losses may be experienced by humans as a result of these options, the figure also demonstrates the trade-offs that are inherent in the standard-setting process. By controlling stressors to the aquatic ecosystem (represented by the lines between Land Uses/Sources and Stressors), a management option should improve certain ecosystem services, resulting in gains to individuals who value these services.

For watershed groups not involved in the standard-setting process, the symbols related to water quality criteria and designated use attainment (Figure 8.2 and Figure 8.3) could be left off the conceptual model or presented separately as additional information for the stakeholders. However, because watershed groups will likely have water quality goals, alternative approaches to illustrate water quality will need to be created for the conceptual models.

At the same time, however, management options that control stressors may impose losses on certain individuals. Some of these losses will result from the *direct* costs associated with controls (e.g., capital and operating costs for effluent treatment systems). Other losses will result from *indirect* costs, which are the value of foregone opportunities (i.e., "opportunity" costs). For example, restrictions on agricultural land uses will generally result in fewer goods being available from agricultural

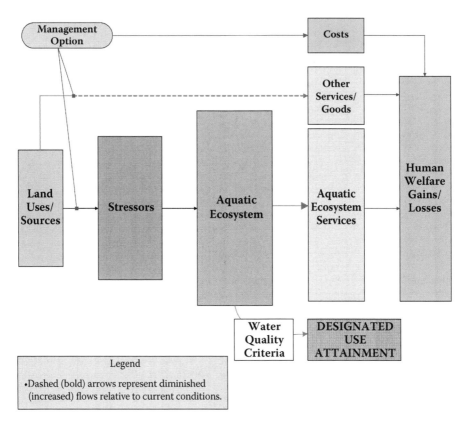

FIGURE 8.3 Effects of management options on aquatic ecosystem services and human well-being.

production. These are important economic endpoints that could be substantial for stream restoration activities.

The economic endpoints must be sensitive to the management alternatives to be considered a gain or loss. If they are not and remain constant, the endpoints drop out of the conceptual models. From Figure 8.3, the key economic endpoints are determined by the ecosystem services, other goods and services, and costs. As discussed in Chapters 1 and 2, an important objective of economic analyses is to quantify the changes in these endpoints. Therefore, the conceptual models link changes to ecosystems, from management alternatives, to changes in human welfare (USEPA 2000).

Figure 8.2 and Figure 8.3 also show how these considerations are related to the attainment of designated uses. Use attainment is ultimately determined by comparing observed water quality (or related conditions) in the aquatic ecosystem with the relevant water quality criteria. Without a management option in place (Figure 8.2), water quality may well be degraded to the point at which specific criteria are not met and the corresponding designated uses are not attained. Once an option is implemented (Figure 8.3), water quality may improve, meeting the specific criteria and attaining the designated use.

STAGES FOR DEVELOPING EXPANDED CONCEPTUAL DIAGRAMS

Applying the general framework outlined to evaluate specific WQS conditions requires gathering and organizing several types of information, first to characterize baseline conditions (based on Figure 8.2) and then to characterize the effects of alternative management options (based on Figure 8.3). The following steps are recommended for these two development stages:

To characterize baseline or current conditions

- list the main ecosystem components and functions that are or could be affected
- list and describe the activities (land uses or sources) in and around the water body that affect or could affect water body integrity
- list the main stressors associated with each activity or source
- identify and show how these stressors are expected or known to enter and impair the ecosystem components and functions
- list the services and goods that are or could be derived from the affected aquatic ecosystems as well as from the land uses and sources
- list the designated uses for the affected water body and, in particular, identify the uses not being attained (or specific water quality goals not being met)
- identify the ecosystem services (and other goods and services) that are or would be primarily affected by the identified land uses, sources, and stressors

To characterize effects of management alternatives

- list the management alternatives that will help to attain designated uses or specific water quality goals set by the watershed group
- determine the types of costs (including opportunity costs) incurred by implementing the management alternatives
- identify and show how the management alternatives will affect the sources or land uses and how they will alter the impacts of stressors on the ecosystem
- identify and show how the management alternatives will strengthen or weaken different ecosystem services (and other goods and service flows)
- identify and show how the management alternatives will positively or negatively affect different aspects of human welfare

To further illustrate how the general approach might work, we present a conceptual model example that uses acid mine drainage (AMD) as the stressor. This example is described through a case study of a hypothetical situation, but one that is based on existing AMD situations. As part of the communication process, we have also created "maps" that precede the conceptual model illustration. The maps reveal the physical aspects of the particular situation; we believe the associated narrative and maps help to grasp the situations.

HYPOTHETICAL ACID MINE DRAINAGE CASE STUDY

In the early 1900s, parts of Pennsylvania and West Virginia prospered with the extraction of coal. Since that time, coal mining has declined, and adverse environmental impacts have increased (especially from abandoned mine lands).

A tributary to a popular recreational river is a major source of AMD. The drainage from the surface mining and tailings has low pH from contact with pyrite (an iron sulfide) and has elevated the levels of metals; AMD can contaminate drinking water sources, eliminate habitat and aquatic life, and corrode infrastructures like bridges. Designated uses for the tributary are aquatic life support, secondary (i.e., incidental) water contact recreation, and agricultural water supply; designated uses for the river are aquatic life support (specifically for warm-water species such as smallmouth bass), primary (i.e., full) contact recreation, and agricultural water supply. These designated uses are not being met in some stretches of both the tributary and the river.

The tributary is about 7 miles long and receives AMD from surface runoff linked to abandoned mine lands and mine tailings (this occurs 3 miles from the head waters). Two seeps emanating from the tailings are visible. Aquatic life, like fish and salamanders, is not found in the tributary after the drainage enters it.

The river, which has many activities that are affected by the AMD, is considered dead for 8 miles after the tributary enters it. However, the tributary is not the only cause of degradation in the river. Several smaller nonpoint sources of AMD also directly discharge into the river along this 8-mile stretch and contribute to poor water quality in this part of the river. Although the riparian habitat is of good quality and other wildlife is abundant, aluminum concentrations prevent any fish population from becoming established in this part of the river.

> The WQS regulations lists six different factors states may use to demonstrate that attaining a use is not feasible (40CFR 131.10(g)). The economic factor (Factor 6) states that "controls more stringent than minimum technology requirements (as specified in sections 301(b) and 306 of the CWA) would result in substantial and widespread economic and social impact. The Interim Economic Guidance for Water Quality Standards: Workbook (USEPA 1995) defines financial ratios to determine whether impacts are substantial and it identifies socioeconomic indicators that should be considered when assessing whether impacts are widespread. USEPA (2007b) provides a framework that includes ecological effects into these types of decisions.

A number of activities and land uses occur in the vicinity of the river and tributary. The river is known for its white-water rafting and kayaking. Hiking, mountain biking, and picnicking are popular around both the river and the tributary, especially along a recently completed rail trail that follows the river and crosses the

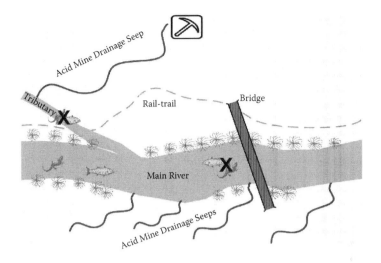

FIGURE 8.4   Current condition map.

tributary. Most recreationists are not from the local area. The tributary and river are not a source for drinking water, but the tributary (above the AMD) supports some stock watering. Forests and pastures are the primary land uses in the watershed. Ten houses are near the tributary; 300 houses are within 5 miles of the impaired river. Figure 8.4 illustrates the current or baseline condition in map form.

The Xs on the fish and salamander represent the fact that the aquatic life use is not met in these stretches of the river and tributary. The pick axe represents the abandoned mine lands from which the AMD flows into the tributary and river. Using the conceptual model, we connect the sources of AMD to the effects on ecosystems, including their services (Figure 8.5).

The conceptual model in Figure 8.5 was constructed using the basic model in Figure 8.2, adding specific details and linkages for the AMD example. The sources of abandoned mine lands and surface runoff produce stressors. These stressors affect infrastructure, like bridges, and the tributary and river ecosystems. Direct and indirect impacts affect the production of aquatic ecosystem services. For example, recreational fishing cannot exist in the current condition, so it is not highlighted like the other aquatic ecosystem services. Designated uses for both the tributary and the river are crossed out because they cannot be met given the current water quality.

In addition to considering the imposition of total maximum daily loads (TMDLs; see Glossary) for aluminum, iron, and pH, the state has also conducted a UAA for the tributary and part of the river. In the UAA, the state estimated the costs of restoring both the entire segment of the tributary and the affected portion of the river. Based on the economic factor for UAAs, the state determined that they cannot afford to do all the restoration. In addition, the state ascertained that the tributary produces more AMD than the combined discharges from the other nonpoint sources that directly affect the river. The results indicate that the costs of restoring the tributary would be considerably less than controlling the nonpoint sources along the river.

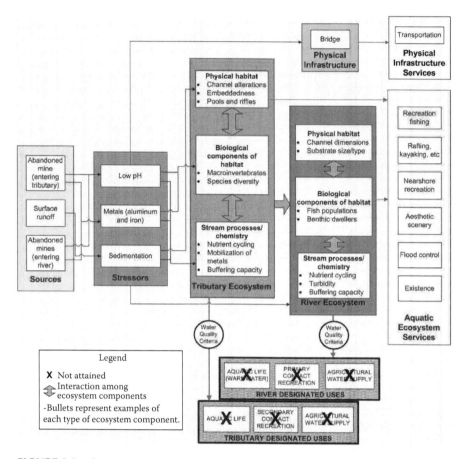

FIGURE 8.5   Conceptual model of current conditions.

Based on the results, the state has decided to focus on restoring the tributary. Several methods are available to raise the pH of AMD-contaminated water from the tributary; however, the two most promising methods available to mitigate the effects of AMD in the above-mentioned reach are (1) a limestone channel and (2) constructed wetlands.

The first option involves installing an open limestone channel and a settling pond. A small dam would be created before the seeps enter the channel to trap sediment and other debris. The channel includes a limestone sand liner and limestone rocks. With a pH of 4.0, the water flows through the channel to a settling basin. The treatment is expected to last 20 years and measurable differences in the tributary are likely to begin in year 1. However, there is a 10% chance the system will fail to meet the tributary's water quality goals. This option is expected to cost $100,000, including excavation costs and land costs. Maintenance costs will be about $2,000 per year after year 1. After 10 years, new limestone rock may be necessary at additional cost.

The second option is to construct a series of wetlands on a large tract of land, just before the seeps enter the tributary, which could be built to reduce metals and

AMD. First, after flowing into a settling pond, a smaller wetland reduces flow and causes metals to precipitate out. The larger wetland further reduces flow velocity and metals; a final settling pond is used for any remaining precipitation. To adequately reduce pH, it will be necessary to augment this system with additional alkalinity. The chance of complete failure of this type of system is about 30%. These wetlands are expected to last 20 years, but the noticeable differences in the tributary will only begin to occur starting in year 3. Costs of the wetlands are expected to be $200,000, including land purchases and maintenance costs of $500 per year after year 1.

Both management options will eventually allow the tributary to support aquatic life, but few anglers will be able to use this resource due to private property restrictions. Restoration of the tributary will improve the overall aesthetic value of whitewater rafting and kayaking in the river; an additional 1,000 person-days per year of kayaking (e.g., 250 individuals kayaking an additional 4 days) are expected. Both options will also allow part of the impaired portion of the river to meet its designated use for warm-water aquatic life; however, the other nonpoint sources of AMD on the river will continue to affect the river quality beyond those restored miles. Property values are expected to increase slightly with either alternative, although there may be an issue related to wide construction "rights of way" for either the limestone channel or wetlands. There is a small possibility that new construction of houses and cabins could occur with the restoration.

With the limestone channel, no additional wildlife habitat will be created near the tributary. However, the limestone channel will provide more buffer capacity for the river than the wetlands. The river is expected to meet its warm-water aquatic life use for 3 miles after the tributary enters it if the limestone channel is used and only 2 miles for the series of wetlands. Given the popularity of fishing in the area, the additional 3 miles that meet warm-water aquatic life could create approximately 200 person-days of recreational fishing. Fewer person-days of recreational fishing on the river are expected if wetlands are constructed.

In contrast to the limestone channel, the constructed wetlands will create additional wildlife habitat, which will enhance recreational and other activities near the tributary. In particular, users of the rail trail (hikers, bikers, and picnickers) will benefit from the new ecological resource, and as a result, an additional 750 person-days per year of hiking, biking, and picnicking are expected. In addition, the wetlands are expected to reduce sedimentation in the tributary and reduce flood potential through surface water storage.

Figure 8.6 maps the limestone channel. Based on the description, we mapped the limestone channel and settling pond onto the illustration and removed the X's from the designated uses for the tributary to show that aquatic life is attained. The limestone channel also allows the river to meet its designated uses for 3 miles below the tributary.

With the first alternative for restoring the tributary, the limestone channel controls only two of the three sources of AMD to the river (not abandoned mines entering river, i.e., in Figure 8.7). Because the limestone channel does some environmental good, the positive flows (represented by arrows) to the ecosystem services are strengthened. For example, as a result of this possible restoration alternative,

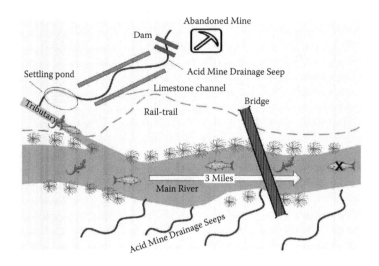

FIGURE 8.6 First restoration alternative for tributary, limestone channel.

recreational fishing would once again become feasible 3 miles downstream on the river. The trade-offs are illustrated in the Human Welfare Gains/Losses box on the far right, which is linked directly to aquatic ecosystem services and the construction and maintenance costs for the limestone channel. The gains and losses in each of the human welfare categories (e.g., recreation values) are represented by + (plus) or − (minus) signs.

Figure 8.8 and Figure 8.9 depict the management option for creating a wetland area. Notice that this conceptual model is similar in many respects to the conceptual model developed for the limestone channel, allowing for easy comparison of various components. The primary difference is that we assume that the wetland supports additional ecosystem services that the limestone channel does not. Therefore, we include the arrow that goes from the constructed wetland option directly to the ecosystem services box. This represents the additional wildlife habitat and reduction in flood potential produced by the wetland area. However, the wetland also produces one less restored river mile compared to the limestone channel. These differences are important when examining the trade-offs of the management alternatives. In addition, the cost of the wetland option is more than the limestone channel, as indicated by the two minus signs in Figure 8.9 as compared to one minus sign in Figure 8.7. It is important to note that the number of plus or minus symbols shown represents the ranking of the management options (limestone channel vs. constructed wetland) regarding their effect on the welfare category. The number of plus or minus signs should *not* be interpreted or used to compare changes across human welfare categories within a single management option (e.g., two minus signs on the disposable income category for the wetland options is not meant to be compared to the number of plus signs on the recreation or residential values for that option).

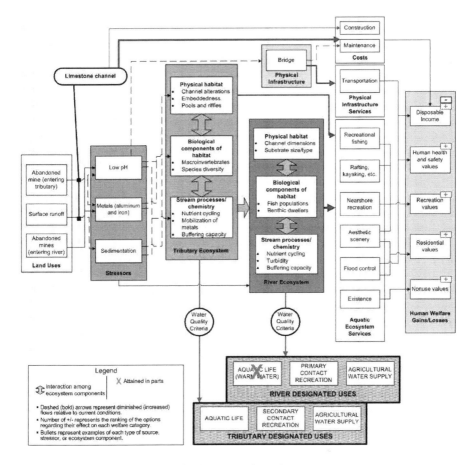

FIGURE 8.7 Conceptual model for limestone channel.

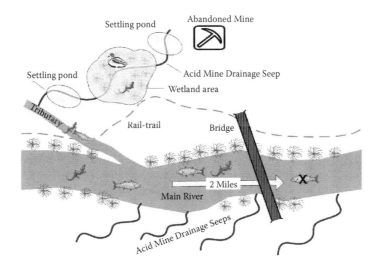

FIGURE 8.8 Map of constructed wetland to control acid mine drainage.

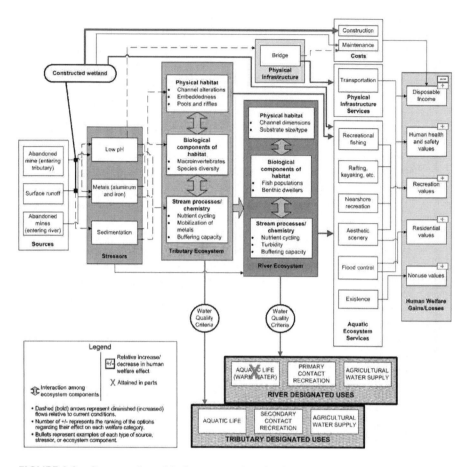

FIGURE 8.9 Conceptual model of constructed wetland restoration alternative.

Without using one or more of the valuation approaches described in the previous chapters, we can only guess what the plus and minus signs mean in terms of actual dollar values. However, given some basic information, we can state something about the relative size of the plus and minus signs. That is, generally speaking, the plus and minus signs should be used to compare the change in one category from one management alternative to the same category in the other alternative. For example, we see that there is one plus sign in recreation values for the limestone channel and one plus sign for recreation values in the wetlands alternative. Compared to the wetland option, the limestone channel would create more of an improvement in recreation services on the river; however, it would not provide the recreation services from the wetland itself. Without further study, we cannot say whether these recreation differences are offsetting or one or the other option would be recreationally preferred. The wetland option would however be more costly than the limestone channel, which is represented by the double negative signs on disposable income for this option, compared to a single negative sign for the limestone channel.

Additional information is required to complete the conceptual models. We also include what data are available for analyses and provide assumptions to make the conceptual models clear.

STAKEHOLDERS

Recreationists, watershed group, homeowners, and State Department of Environmental Protection (DEP).

DATA AVAILABLE

DEP collects data on certain water bodies that are impaired and require TMDLs. A university has studies related to AMD in the area, including methods and cost-effectiveness. A watershed group has developed a watershed plan that describes issues related to AMD throughout the watershed, not just the tributary and specific stretch of the river.

ADDITIONAL ASSUMPTIONS

- The only two significant sources of stressors on the tributary and the 8-mile portion of the river are abandoned mines and surface runoff (sedimentation).
- Healthy riparian habitat in the tributary and river help to control surface runoff and prevent flooding downstream.
- The only significant service provided by the bridge is transportation, and the main cost associated with corrosion is more frequent maintenance.
- As long as the two management options do not fail, they will both allow for all designated uses on the tributary and river to be met, with the exception of warm-water aquatic life use in the river, which will still be affected by other sources of AMD.

For the economic endpoints, measuring specific costs and changes in welfare requires additional information that could be difficult to acquire. For example, estimating the additional recreational days could require specific models developed by experts; costs of different management alternatives may need to be estimated by engineers. Nevertheless, more easily acquired knowledge regarding the *relative* size of the plus and minus signs might be sufficient to support a decision.

Watershed groups will have to determine what specific information will be needed for their audiences. As compared to Figure 8.9, Figure 8.10 includes less detail regarding the complex relationships among sources, stressors, aquatic ecosystems, ecosystem services, and human welfare changes. This intermediate-level diagram may be useful as a tool for communicating the water quality management problem to the broader public in the affected community. Parts of these diagrams could, for example, be used in public meetings as a way of walking the community

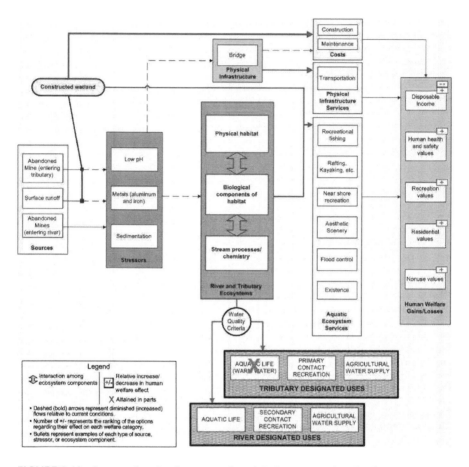

FIGURE 8.10   Intermediate-level conceptual model of constructed wetland.

through the issues and trade-offs involved in the AMD sources and as a way of eliciting further feedback from the public.

SUMMARY

We have introduced conceptual models as an approach to illustrate how changes to water quality can affect both economic and ecological endpoints. A more comprehensive view of the relevant impacts and trade-offs can be achieved by considering how ecosystem services are affected by alternative management options. Important insights can be gained by identifying and considering the full range of socioeconomic endpoints related to ecosystem services even when it is not feasible to conduct a detailed, quantitative benefit-cost analysis. Several examples of these endpoints are shown as "human welfare gains/losses" in the conceptual diagram. As this terminology implies, any change in human well-being resulting from management alternatives or WQS changes should be interpreted as a relevant socioeconomic endpoint.

Changes to aquatic ecosystem services can affect human well-being in a variety of ways. Some of these effects will have direct monetary or market implications for individuals. For example, several services provided by water resources, such as commercial fishing, energy supply, and agricultural water supply, directly support market activities. As a result, changes in these services can affect both producers and consumers by changing the costs of production, prices, incomes, and employment related to these activities.

Some aquatic ecosystem services have less-direct but equally relevant monetary or market implications. For example, flood control services help prevent financial losses associated with property damage, and aesthetic services for nearshore residents are reflected in housing prices and property values. Changes to these services can therefore also have impacts on prices, incomes, and employment (in these cases, mostly related to property markets and ownership). These socioeconomic endpoints also deserve consideration.

Other aquatic ecosystem services have little or no connection to markets or incomes; nevertheless, they are still valued by individuals and contribute to their well-being. Recreational services are a prime example. If, for instance, services from recreational fishing, boating, swimming, or other activities are affected by changes in water quality, these changes will not necessarily affect prices, incomes, or employment in any market. However, the absence of a direct monetary effect on individuals does not imply that there is no socioeconomic effect. In these cases, the relevant endpoint is the change in enjoyment individuals derive from their recreational activities. For example, reducing the effects of stressors on aquatic ecosystems is shown to enhance their recreational services and to provide more value to recreational users of the resources.

Several other categories of ecosystem services have similar "nonmarket" characteristics. For example, in many cases, changes to aesthetic services or changes to services derived from cultural and subsistence activities will not have observable effects on prices or incomes. Again, despite the lack of a direct monetary impact, the change in individuals' enjoyment of these activities represents a potentially important socioeconomic endpoint to be considered.

One category of ecosystem services is unique because it is not derived from any specific use or market related to the aquatic resource—nonuse/existence services. The effects of these services on human well-being are less tangible than other services and certainly more difficult to measure, but they may nonetheless be significant. The argument for considering these services is that individuals may well value protecting the existence and quality of natural resources that they never expect to use in any way. These values are likely to be particularly strong for aquatic resources that are unique, threatened, or endangered. Regardless of the motivation for nonuse values, they represent another potentially important socioeconomic endpoint to consider as part of setting or modifying WQS, and for this reason they are included as potential human welfare gains in this conceptual model.

From the discussion here, it should be apparent that, in addition to the involvement of individuals who know and use the aquatic ecosystems in question, various types of experts are needed for the development of the conceptual models. It would be unrealistic to expect that one expert is completely familiar with all of the endpoints for any particular watershed problem. In fact, the development of the conceptual model helps create an interdisciplinary approach for understanding the ecological and economic effects of management alternatives or changes in pollution levels. If the conceptual model itself does not provide sufficient information to enable a decision and a quantitative analysis is required, the conceptual model provides an interdisciplinary, analytic road map. Each step along the path may require some expertise that may or may not be easily accessible. The watershed groups will have to consider what specific expertise is needed by considering the important endpoints in the model. We suggest considering your state's environmental agency and universities as well as local community members as possible sources of information; additional resources for developing conceptual models, including examples, can be found in the USEPA references (1998, 2000, 2002, 2007a,b).

ACKNOWLEDGMENTS

We would like to thank Dan Sweeney and Tim Connor for their participation in the 2006 USEPA workshop, "Weighing Ecological Risks, Costs, and Benefits in Use-Attainment Decisions," which was used to revise the summary text and conceptual models in this chapter.

REFERENCES

Daily, G.C. (Ed.). 1997. *Nature's Services: Societal Dependence on Natural Ecosystems.* Washington, DC: Island Press.

Millennium Ecosystem Assessment. 2005. *Ecosystems and Human Well-Being: A Framework for Assessment.* Washington, DC: World Resources Institute.

U.S. Environmental Protection Agency (USEPA). 1995. *Interim Economic Guidance for Water Quality Standards: Workbook.* EPA-823-B-95-002. Washington, DC: Office of Water.

U.S. Environmental Protection Agency (USEPA). 1998. *Guidelines for Ecological Risk Assessment.* EPA-630-R-95-002F. Washington, DC: Risk Assessment Forum.

U.S. Environmental Protection Agency (USEPA). 2000. *Assessing the Neglected Ecological Benefits of Wetland Management Practices: A Resource Book.* Washington, DC: Office of Water.

U.S. Environmental Protection Agency (USEPA). 2002. *A Framework for the Economic Assessment of Ecological Benefits.* Washington, DC: Office of Science Policy, Social Sciences Discussion Workgroup. http://www.epa.gov/OSA/spc/pdfs/feaeb3.pdf (accessed March 13, 2007).

U.S. Environmental Protection Agency (USEPA). 2006. *Ecological Benefits Assessment Strategic Plan.* EPA-240-R-06-001. Washington, DC: Office of Administrator.

U.S. Environmental Protection Agency (USEPA). 2007a. *Causal Analysis/Diagnosis Decision Information System (CADDIS). Conceptual Model Library.* Cincinnati, OH: Office of Research and Development. http://cfpub.epa.gov/caddis/examples.cfm?Section=26 (accessed March 13, 2007).

U.S. Environmental Protection Agency (USEPA). 2007b. *A Framework for Evaluating Trade-Offs and Incorporating Community Preferences in Use Attainment and Related Water Quality Decision-Making.* External Review Draft. NCEA-C-1658. Cincinnati, OH: Office of Research and Development.

9 Local Economic Benefits of Restoring Deckers Creek
A Preliminary Analysis

Alyse Schrecongost and Evan Hansen

CONTENTS

INTRODUCTION

This chapter uses concepts and methods presented in the book to justify the use of public funds to restore Deckers Creek in West Virginia. It describes a simple but useful example for watershed groups to consider for public outreach and advocacy.

PROBLEM STATEMENT AND BACKGROUND

The Deckers Creek watershed comprises 64 square miles in Preston and Monongalia counties, West Virginia. The watershed's largest city, Morgantown, lies at Deckers Creek's confluence with the Monongahela River and is home to West Virginia University. River-focused development and a popular 19-mile rail trail that parallels Deckers Creek and its tributary Kanes Creek have helped increase the community's interest in the remediation of Deckers Creek. The local Convention and Visitors Bureau promotes the city as a hub for outdoor recreation and nature-based tourism. While Deckers Creek was once used as a de facto waste disposal system, it is now emerging as a potential centerpiece of economic development.

A "state of the creek" report conducted annually by the Friends of Deckers Creek (FODC), a very active local watershed association, documents trends in water chemistry, fish communities, and benthic macroinvertebrates at 13 sites across the watershed. According to these monitoring data, the entire watershed is not completely dead, but significant portions are impaired by acid mine drainage (AMD) pollution (Christ 2004, 2005). Most of this pollution is caused by old coal mines abandoned before the 1977 surface mining law.

FODC's short-term goal is to work with agencies to install AMD-remediation projects so that no streams in the watershed remain chemically impaired. The organization's long-term objectives are to reestablish a healthy fishery and to make a clean Deckers Creek a centerpiece of the community and a point of pride. The original report on which this chapter is based was part of a focused advocacy effort to encourage local government officials to commit public funds for cleaning up one of the worst polluters of the Deckers Creek Corridor: Richard Mine. The discharge from this mine is responsible for virtually all of the AMD pollution as Deckers Creek flows through Morgantown.

Originally, this economic benefit analysis was conducted specifically to educate community leaders about the link between local economic benefits and restoration of the Deckers Creek watershed as a valuable economic asset. In this case, FODC had secured funds for installing an adequate pollution treatment project at Richard Mine but still needed the funds to operate and manage the restoration system annually. This chapter evaluates economic value already generated by related upstream restoration projects in the Deckers watershed, plus it estimates economic benefits anticipated from cleaning up Richard Mine specifically in an attempt to justify the remaining public investment needed.

This type of analysis is an advocacy and public opinion tool. This approach is not appropriate for prioritizing a list of restoration projects by comparing different projects' cost-to-benefit ratios because additional and more precise methods would be necessary (e.g., discounting; see Chapter 1). It is possible that the Richard Mine project may generate the most economic benefit per dollar invested; other factors, however, were considered when FODC wrote its watershed-based plan to tackle some smaller problems first. These factors included ease of access to polluting properties and the priorities and plans of partner organizations. Also considered was project size relative to the experience and capacity of the organization; smaller passive treatment systems were simply more affordable and manageable for the nascent organization. Also related was the community's increased awareness of the pollution in Deckers Creek. Prior to building the rail trail and without the years of FODC community education efforts, there may not have been sufficient community support for Deckers Creek to back the costly Richard Mine cleanup project.

With FODC and partner agencies' demonstrated successes on smaller projects in the watershed, extensive community education, and the increased public familiarity with the creek via the rail trail, the public's demand (and, therefore, their willingness to pay [WTP]) for creek remediation almost certainly increased accordingly. Attempting to tackle the creek's biggest and most expensive source of pollution without this increased public familiarity with the creek would have been challenging to impossible. Furthermore, tackling some of the smaller projects provided the organization with valuable experience and skills for managing a larger project.

BENEFITS

Environmental restoration, particularly if combined with asset management, generates three primary categories of local economic benefits (summarized in Table 9.1). The first category includes the benefits that result from one-time restoration spending at local businesses (e.g., engineering and construction, surveying, environmental testing, and nurseries). Chapter 1 calls this economic impact. This spending is considered a local economic impact rather than a cost if funding originates outside the local economy. For Deckers, federal and state dollars financed most of the local restoration. Because the funds came into the local economy from outside specifically and only for Deckers Creek restoration, the study does not have to compare the value of investing in Deckers Creek against some other investment project (a *trade-off*) to find the opportunity costs of using the money for another local project.

The second category is the flow of benefits to the local economy that is generated continually by the restored stream (e.g., urban recreation opportunities, increased tourism, and increased property values). For example, boaters anxiously await the opportunity to kayak in clean, safe water through the challenging stretch of Deckers Creek called the Miracle Mile, well known among local and regional boaters. Benefit flows include financial benefit flows that result in dollars circulating in the economy and nonmarket benefit flows that encompass nonuse and indirect use values, existence, and bequest values of the stream (Loomis 2000). Estimating these nonmarket economic benefits requires an evaluation of people's WTP for changes in the creek using revealed or stated preference methods. Chapter 3 describes value estimates for the creek derived from the stated choice method (SCM)—a survey that elicited people's stated preferences for their WTP for a fully restored Deckers Creek. Finally, increased property values, estimated by using benefit transfer for a hedonic analysis, are another source of a flow of benefits from the asset of a restored Deckers Creek.

The third category of benefits can be considered "costs avoided" and accrue from halting or reversing degradation trends (e.g., lower water treatment and health care costs and reduced flooding and sedimentation). Avoided costs can also include the creation of space that is attractive for development on land that is already served by

TABLE 9.1
Economic Benefits of Deckers Creek Restoration Efforts

Type of Benefit	Description	Estimation Method	Potential Benefits (Million $) One Time	Annual
Restoration spending (one-time expenditure)				
Local impact of expenditure of external project funds	Federal and state dollars attracted to Preston and Monongalia County economies	IMPLAN input-output model of Monongalia and Preston counties	14.16	N/A
Economic benefits from a restored Deckers Creek (flow of benefits annually)				
Nonmarket values	Estimate of the value people hold for goods (nonuse values of creek heritage, beauty, existence, etc.); usually a perceived financial estimate of ethical value	Stated choice method: survey of various groups' willingness to pay higher utility bills to finance improvements to creek	N/A	1.9
Local expenditure by increased visitors	Local spending by visitors attracted to new or improved opportunities on and around Deckers Creek	NRCS estimate	N/A	1.16
Increased property values (one time)	Improved value of streamside properties and nearby neighborhoods	Potential property value gains for streamside properties along Deckers Creek downstream of I-68 (streamside neighborhood properties excluded)	0.95	N/A
Costs (damages) avoided (annual or one-time)				
Costs avoided by changing status quo	Avoiding further reduction in visitor enjoyment and expenditures, reducing associated health costs, and improving streams' abilities to resist and recover from future disturbances	Not estimated	N/A	N/A

N/A, not applicable.

city infrastructure but is otherwise degraded; this is important particularly in cities that are experiencing rapid and expensive infrastructure expansion demands such as Morgantown. In addition to these three categories, many other valuable benefits are generated that cannot be reliably estimated.

It is important to note that the estimations described can sometimes end up counting some of the same or similar benefits in two different ways. As a result, these benefits cannot simply be added together without artificially inflating the estimate of total benefits; this is called *double counting*. It may be useful for a watershed group, however, to use different types of estimation methods for the same benefits if they will resonate with different audiences. In this example, estimating property value increases separately through hedonic analysis sent a clear message to local government leaders and other interested parties that efforts to improve certain neighborhoods could benefit from investing in environmental amenities.

The next section provides an economic impact analysis (first category of benefits): one-time restoration expenditure impacts on the local economy. For this section, the local economy is defined as Monongalia and Preston Counties. IMPLAN (IMpact analysis for PLANning), an input-output model of the local economy, traces expenditure patterns by sector to determine how many times $1 spent in a local business will be respent locally in other sectors before it leaves the local economy. Specific attention is given to the planned remediation of the Richard Mine.

Then, we describe the potential flow of benefits expected from a restored Deckers Creek. Likely economic benefit flows include those related to new or expanded recreation opportunities, increased production of ecological services, aesthetic improvements, wildlife habitat, and others. Potential economic benefits from increased visitor expenditures and increased riparian property values are also discussed in this section.

The final section of this chapter reviews conclusions and suggests opportunities for a more comprehensive economic analysis of the Deckers Creek watershed. Among those benefits not addressed by this study are benefits from community environmental education, restoration of ecological assets, and the entire third category of benefits identified: the benefits of costs avoided.

ECONOMIC IMPACT ANALYSIS OF DECKERS CREEK RESTORATION PROJECTS

AMD remediation projects require either passive or active treatment systems. Passive systems are installed, and remediation occurs largely without further intervention. More severe AMD sites require active systems that require continual, often expensive, operation and maintenance (O&M) activities.

Since its beginning in 1995, FODC has already helped to attract almost $2 million in direct external funds through July 2005 for passive system remediation in the Deckers Creek watershed (Table 9.2). Using the IMPLAN model of the local economy, that spending on passive treatment projects is estimated to have generated $2.52 million in benefits to local businesses and families. More than $8 million of additional funding is expected in the next few years.[1]

TABLE 9.2
Deckers Creek Watershed Restoration Spending as of July 2005 ($)[a]

Funding Source	Spent	Committed	Pending Approval	Total
Abandoned Mine Land Trust Fund	1,595,000	3,205,000		4,800,000
NRCS		4,800,000		4,800,000
Clean Water Act Section 319	188,000	58,000	238,000	484,000
Office of Surface Mining WCAP	180,000	94,000	158,000	432,000
Total	1,964,000	8,157,000	395,000	10,516,000

[a] Totals may not equal sum of funding sources due to rounding.

IMPLAN estimates how expenditures have an impact on an economy by tracking how project funds recirculate through the local economy for the purchase of locally produced inputs and provision of local employment benefits.[2] For example, a dollar spent in stream restoration circulates in the local economy approximately 1.28 times—this is called the *multiplier,* and it varies depending on the location and sector of the spending.[3] In the case of Deckers Creek, remediation costs are paid by external (state and federal) funds brought into the two-county area, so the impact is $1.28 of local economic activity.[4] A local dollar spent in remediation would have generated $0.28 of additional local economic activity beyond that dollar spent.

Passive Treatment Remediation

Efforts of FODC and partner agencies as of 2005 have resulted in the partial or full remediation of at least four important AMD sites in the watershed.[5] The economic impact benefits from passive treatment expenditures were estimated by applying the IMPLAN-generated multiplier (1.28) to total passive treatment expenditures to date ($1.96 million).[6] So, total economic impact benefits in Monongalia and Preston counties from the installation of passive treatment systems in the Deckers watershed amounted to the $1.96 million direct dollars expended plus $409,855 that recirculated through the economy (a total of $2.52 million). This expenditure was estimated to have generated an additional 3.6 employment years, yielding over $143,000 in labor compensation (a total of $879,996 in employment compensation).[7]

Active Treatment Remediation: Richard Mine

Richard Mine is the most significant pollution source remaining on Deckers Creek (Figure 9.1). Economically, its pollution impact is particularly significant because it visibly affects the stream along the Deckers Creek Trail and is almost single-handedly responsible for the AMD pollution visible in Morgantown. For most local residents, bright orange rocks and milky flows are telltale signs of an AMD-impaired stream.

Because of its scale, effectively treating the polluted discharge from Richard requires the costly installation of an active treatment system. Active systems require

FIGURE 9.1   The Richard Mine discharges 200 gallons per minute of acid mine drainage directly into Deckers Creek.

significant ongoing O&M costs. While external funds are available for fixed costs, including site preparation, equipment purchase and installation, and land acquisition, release of these funds requires a secured source of funding for anticipated O&M costs into the indefinite future. FODC estimated that the O&M costs it is seeking from local government will total $88,140 annually.

Installation Costs

Using an IMPLAN multiplier[8] of 1.17 for the active treatment system, the one-time project installation expenditure alone ($144,887) would generate a $169,789 benefit to the two-county study area, of which $61,490 would be labor compensation. Externally funded land acquisition costs of $250,000 would be added to the final impact, resulting in a total local economic benefit[9] of at least $420,000.

Operation and Maintenance Costs

Predicting benefits of annual O&M costs paid with local funds is more complicated. The annual $88,140 costs could be expected to generate a $108,002 impact locally, and income compensation would account for $36,307.[10] If this O&M funding comes from local sources, it should be considered a cost and compared with benefits forgone (*opportunity costs*) from alternative use of the funds. Arguably, this expenditure should be counted as a full $88,140 in annual costs to the community; it then becomes important to demonstrate that the economic benefits of the Richard Mine cleanup will surpass $88,140. This is done in the section on restored benefits.

TABLE 9.3
**Estimated One-Time Economic Benefits to Monongalia and Preston Counties
from Passive and Active Remediation Projects (million $)**

Expenditure	Total Direct Benefits	Indirect Benefits	Total Impact
Implemented			
Passive treatment and planning	1.96	0.56	2.52
Planned			
Richard Mine active system installation	0.55	0.03	0.58
Future passive treatment systems and other restoration	8.55	2.51	
Total	11.06	3.10	

ANTICIPATED REMEDIATION PROJECTS

AMD treatment systems and other remediation projects are needed on multiple Deckers Creek tributaries as well as on the main stem. Some of these projects have been designed, and external funding has been committed or is pending. Priorities are being set for other projects now. The estimated local economic benefits of these projects are considered in Table 9.3.[11]

AUTHOR'S MIRACLE MILE ANECDOTE

While editing this report, a friend called from Pennsylvania. We debated who would come to visit whom. Coincidentally, my friend explained that he would be more than happy to come to Morgantown if we would clean up Deckers Creek so he could boat the Miracle Mile without risking another trip to the hospital—a treat he experienced on his last visit after accidentally swallowing a bit of creek water in a difficult rapid. As a direct economic result, rather than attracting a visitor and his spending to Morgantown, both of our expenditures for the weekend were made boating in Pennsylvania.

BENEFITS FROM THE RESTORED STREAM

The second category of benefits generated from restoration is the flow of economic benefits that comes from a healthy watershed. If the watershed is impaired, the benefit flow decreases and can even become negative, generating costs to the community rather than benefits. Benefit flows evaluated in this section include the following:

FIGURE 9.2   Friends of Deckers Creek clean up trash along the creekside Old Route 7 Scenic Byway.

(1) nonmarket, quality-of-life benefits; (2) expenditure in the local economy due to increased creek and trail use; and (3) increased streamside property values.

In the case of Deckers Creek, there are multiple sources of stream impairment. The three primary impairments to Deckers Creek are AMD, garbage, and bacteria from human sewage and animals. Correcting one issue will not proportionately improve recreation opportunities because stream use is largely dependent on comprehensive restoration. Fish may return with AMD remediation, but anglers will not choose to fish a stream impaired by sewage and garbage.

Active and passive treatment projects tackle the AMD problem. FODC volunteers have already removed tons of garbage and tires from the creek bed and banks—a significant economic value donated by the organization and its supporters (Figure 9.2). Realizing the full benefits from investment in these two issues will require investment in solving wastewater discharge problems. Wastewater problems are likely to require local investment, but such a project is easier to advocate with the local utility board given the demonstrated benefits of state and federal remediation projects and countless hours of volunteer efforts.

Nonmarket, Quality-of-Life Benefits: Willingness to Pay

Quality-of-life benefits enjoyed by residents from creek restoration are commonly called *nonmarket goods* because there is no formal purchase price for them, but they do hold value. This nonmarket value is considered to be a *flow benefit* because it is enjoyed continually and not just during the one-time act of restoring the stream. Residents' value for an improved quality of life can be estimated with the SCM or surveys designed to estimate their WTP for restoration. This section reviews a study conducted in the Deckers Creek community by West Virginia University professors in agriculture and natural resource economics.[12]

Collins et al. (see Chapter 3) sought to estimate the value of improved scenic benefits (reduced trash and reduced visible effects from AMD), improved angling benefits (restored habitat for fisheries), and safe water-contact recreation benefits (reduced bacteria). Researchers used a survey of Monongalia and Preston County residents, questioning residents about their value for a clean Decker's Creek.

In this study, average WTP for full restoration ranged between $12 (nonanglers) and $16 (anglers) per household per month (see Chapter 3, Table 3.3). Adding these benefits across the watershed population yielded benefits totaling $1.9 million. Restoration of Richard Mine alone would improve two of three creek attributes—aquatic life (Figure 9.3) and scenic values—but not primary contact recreation or odor.[13]

The Collins et al. (2005) study was conducted with current residents; it did not include visitors' values, which could increase the total value of Deckers Creek. The importance of natural amenities and stewardship on homebuyers' location decisions and on young professionals' location decisions should not be underestimated as another important *flow benefit,* although this article did not estimate its value specifically. Many studies have shown that natural and cultural quality-of-life amenities are increasingly important factors in firm location decisions, particularly for the knowledge-based industries of the new economy (Salvesen and Renski 2002). The authors

FIGURE 9.3   Restoration is improving habitat for fish like these found in Deckers Creek.

specifically addressed the unique opportunity for cities in rural regions to attract firms by offering cultural amenities while retaining natural amenities such as clean air, environmental quality, recreation opportunities, and community attitude. As Morgantown works to attract high-tech businesses and retain educated young professionals, protecting and promoting its natural amenities will distinguish it from other cities.

ECONOMIC BENEFITS OF INCREASED RECREATION VISITS

Increasing the use of Deckers Creek can be assumed to increase the number of visitors and frequency of visitor trips. It is also likely to diversify the types of visitors to the creek and its trails. The Natural Resource Conservation Service (NRCS) estimated that a restored Deckers Creek could be expected to increase trail use by 10% annually (NRCS 2000). Based on trail use in 1999 (estimated 60,000 annual visitors), that would mean attracting 6,000 new visitors for boating, rock climbing, wading, fishing, and other water-based recreational activities annually. Boating alone is expected to at least double in the difficult Miracle Mile stretch of the stream and increase fourfold in the easier stretches as health threats are ameliorated.

Benefits from this increased visitation would include the economic impacts of visitor expenditures at local businesses plus the benefits enjoyed by the visitors themselves from being able to use the river and trails. In 1999, NRCS estimated that annual recreation benefits to the local economy would surpass $1.16 million (original estimate indexed to 2005 dollars).[14]

An added benefit may be generated from encouraging increased use of the trail for health and commuting benefits. The Virginia Department of Conservation and Recreation (Center for Watershed Protection 2001) cited the link between increased parks and recreation activity and decreased health care expenditures. The same study also credited increased trail use for commuting for reducing local commuter traffic, a needed benefit in Morgantown.

PROPERTY VALUES

Property values of homes on and near streams are related to the water quality of those streams (see Chapter 4). *Benefit transfer* is the methodology of drawing on findings from similar case study research and applying results to a new situation (see Chapters 5 and 7). Using the cases cited in this section, increased streamside property values in just the Morgantown section of the Deckers Creek watershed could conservatively be estimated to surpass $568,000.

It is important to note that this $568,000 cannot be added to the findings of the Collins et al. (2005) WTP survey, as reviewed in this chapter, for a combined benefit value grand total. SCM survey respondents with riparian properties likely answered questions based on an assumption of possible increased property values, so adding the values would result in counting some of the same benefits twice. This section is simply another way to measure benefits, specifically property value benefits. Watershed groups may opt to take different approaches to measuring the same benefits based on their objectives, their audience, available funding to pay for the benefit study, and other variables. In this case, having a small section on increased riparian property values was important for the purpose and audience of the report. The section underscores the potential for increased property tax revenues for local coffers (the estimated increase stated would provide the county with an estimated $6,000 additional tax dollars annually from just the immediate riparian properties). Perhaps more important, it sends the message that the Deckers Creek riparian neighborhoods, which are currently degraded and problematic for local officials, would be improved by a cleaner creek.

Estimating the impact of water quality changes on property values presents a few significant challenges. The first is the role of homeowners' and homebuyers' awareness of different water quality levels. The decision makers must be able to observe water quality changes either by perceptible changes in odor, clarity, or color or by regularly published test results. In addition, the effect of stream impairment may not necessarily register as a change in property value but rather as a shift in the community's socioeconomic or demographic characteristics. AMD, bacteria-related odors, and garbage in Deckers Creek are all perceptible sources of water quality impairments. The direction of demographic trends in many streamside neighborhoods in Morgantown is arguably still being defined; stream improvements could help riparian communities retain single-family dwellings among the rental properties.

Many studies have examined the relationship between environmental restoration and increased property values (see Table 9.4). Studies can predict how property

TABLE 9.4
Summary of Selected Studies on Property Values and Restoration in Other Watersheds

Study	Restoration Effort	Benefit/Estimated Benefit
Streiner and Loomis (1996)	Bank stabilization and trail construction on urban stream stretches in three California counties	Property value increases of 3–13%
Epp and Al-Ani (1979)	AMD remediation in rural residential area of Pennsylvania	1 point of stream pH improvement, increased property value by 5.9%
Cameron and McConnaha (2005)	Remediation of area superfund site	Neighborhood returned to pre-Superfund site demographics, attracting families with children
Leggett and Bockstael (2000)	Fecal coliform levels (bacteria from sewage) ranged from 4 to 2,300 counts per 100 mL along shores of same waterfront community; study on variation in housing market	Bacteria counts per 100 mL changed property values decreased by 1.5% per additional 100 counts, with value impacts as high as 34.5%
Earnhart (2001)	Long Island Sound restoration of degraded marshes in residential areas	Marsh restoration increased property values by 16.6%
Krysel et al. (2003)	Water and land management around residential lake to improve water quality and clarity	Increased property value for lakeside homes of $423 per frontage foot after remediation improved lake clarity by 3 ft (home with a 40 frontage foot parcel increased by $17,000)

values would improve after restoration based on similar housing markets near pristine streams or lakes. Research can also follow changes in property values throughout the restoration process, tracking actual improvements. *Benefit transfer* applies findings from existing case research in other watersheds to estimate how local property values may change after restoration.

Property Values in the Deckers Creek Watershed

Approximately 85 parcels border Deckers Creek from the edge of Morgantown to the mouth of Deckers Creek, like those on Brockway Avenue pictured in Figure 9.4 (Monongalia County Assessors Office 2005). Of these properties, 62 have assessed property values on file with the Sheriff's Department (Monongalia County Tax Records 2005). The aggregate assessment value of these properties is $4,366,000 (estimated to be 60% of the market value). Given Morgantown's current growth patterns, transferring a conservative property value increase of 13% would generate $946,000 in property value increases.

Using the most conservative number errs further on the side of conservative valuation because of the conservative definition of riparian properties that would benefit. The aggregated property values considered for this estimate include only properties directly bordering the creek; spillover benefits to properties slightly further (even those just across the street) from the creek were not included. In addition, property value benefits would be expected to accrue to the communities of Richard and Dellslow just beyond the Morgantown border.

FIGURE 9.4 Brockway Avenue home along Deckers Creek, which suffers from bacteria-related odors and health threats.

It is worth emphasizing that many of the riparian properties in Dellslow, Richard, and even streamside in Morgantown have not benefited from the exceptional period of economic growth in the area. This is at least in part due to area's polluted history. Given the communities' proximity to two major highways and to downtown Morgantown, growth and property value increases could be significant if the local AMD source were remediated and natural amenities restored, turning an environmental liability into an economic asset.

COSTS AVOIDED

The third category of benefits from stream restoration is the benefit of costs or damages avoided. Estimating costs avoided is beyond the scope of this report. Areas can be identified, however, in which potentially significant costs could be avoided. Health care, community infrastructure, and community development costs are among those that could be reduced if Deckers' water quality improved. The same issue of double counting, however, occurs with this category of benefits if the Collins et al. (2005) results are used.

Health care costs avoided as a result of remediation of bacterial contamination is another type of cost that could be avoided by a full stream remediation. Estimating these benefits would be difficult because estimates of current creek-related health problems are not available.

Additional costs avoided could be analyzed relative to community planning goals. Based on the Cameron and McConnaha (2005) study of pollution and demographic trends (described in Table 9.4), Morgantown planners' interest in fostering mixed-income neighborhoods and in retaining single-family housing throughout neighborhoods like Second Ward and South Park would be advanced by creek restoration.

Finally, environmental restoration and remediation are expensive activities. While external spending may generate economic benefits to the economy in the short run, investing in environmental degradation prevention allows more costly degradation-related problems to be avoided entirely. Investing now in creek restoration can change public awareness and aversion to activities that degrade the creek by encouraging a greater sense of ownership and protection for the creek among enforcement officials and local users. This type of shift in public awareness and attitude can have important and valuable implications for avoiding future costly threats to the creek quality.

FINAL CONCLUSIONS AND RECOMMENDATIONS

This chapter reviewed estimates of local benefits that could be expected from a full remediation of Deckers Creek. Economic benefits have and will accrue to the local community (Monongalia and Preston counties) from one-time investments made to reduce AMD pollution through installation of passive treatment systems (already $2 million). Combined with funds already slated for future watershed remediation ($8.2 million), project spending to restore the watershed will generate $14.16 million in economic benefits to local businesses and workers.

The nonmarket benefits that local residents will experience from improvements related to increased opportunities for fishing, swimming, and passive enjoyment of a restored Deckers Creek are estimated at between $1.02 and $1.9 million annually

TABLE 9.5

Categories of Benefits from Richard Mine Remediation that Justify Annual Richard Mine Remediation Costs (million $)

Annual Local Benefits		Annual Local Costs	
$1.02	Quality-of-life economic benefits	$0.088	Operation and maintenance costs
$1.16	Recreation spending	—	—

(depending on degree and nature of restoration). Finally, increased recreation-related expenditures are estimated to generate an additional $1.16 million in local economic benefits annually (Table 9.5). We estimate that riparian property owners could conservatively see a 13% increase in their property values, although we cannot directly add these benefits to the estimated nonmarket benefits because of the potential that some portion of those values might be double counted.

Additional benefits have and will continue to accrue from FODC board and volunteer hours donated for events such as garbage removal activities, public education about creek-related environmental issues, fundraisers, and advocacy. As well, costs avoided were not estimated by this study. Restoration not only increases the stream resistance to and resilience from unanticipated natural disasters like severe flooding and drought, but also increases the population's interest in protecting the creek from harmful activities in the future—essentially decreasing people's willingness to accept future harm in exchange for financial compensation. Finally, reductions in AMD and garbage pollution increase pressure on the local wastewater utility to address the only remaining impairment of sewage. Understanding the scale of these benefits warrants additional research, but they can be assumed to contribute significantly to building a vibrant and sustainable local economy.

This chapter illustrated the value of environmental benefit analysis as an advocacy tool. Detailed data and analysis would be needed for combining the benefits and costs for prioritizing projects or choosing among various remediation strategies. For example, *discounting,* touched on in Chapter 1, would play an important role in the analysis if the watershed group would want to combine the benefit and cost estimates. Effectively communicating the link between economic and environmental health to key decision makers may actually increase the value of the environmental good; their increased awareness of and appreciation for Deckers Creek as a community asset should increase how much they value the good and thus increase their WTP for restoring and protecting it.

Findings in this chapter were presented to city council at a public and televised meeting and to the president of the county Chamber of Commerce. Reports were printed and delivered to the county commission and other decision makers in the community. As a result of these and other advocacy efforts, FODC now has the necessary commitments to cover most O&M costs, which allows previously allocated funding for capital costs to be spent on a treatment system for the Richard Mine.

ACKNOWLEDGMENTS

We would like to thank the Friends of Deckers Creek for sponsoring the study and Executive Director Martin Christ for providing data, direction, and editorial support. Thanks to Randy Childs of the West Virginia Bureau of Business and Economic Research, Dr. Alan Collins of the West Virginia University Agricultural and Resource Economics Program, and Pamela Yost of the Natural Resource Conservation Service for their guidance and comments. We would also like to thank the Minnesota IMPLAN Group (MIG) for donating the IMPLAN software and data sets for this study.

NOTES

1. Funds spent as of July 2005 included the following projects: Kanes Creek South, Dillan Creek, Elkins Coal and Coke, Slabcamp Run 2, Deckers Creek Doser and Limestone Fines Study, and Deckers Creek Watershed Based Plan. Funds committed as of July 2005 included the following projects: the rest of the projects described by NRCS (2000) and Valley Point 12. Funds pending approval include the following projects: Valley Highwall 3 and Kanes Creek South Site 1. AML Trust Fund figures are from the Abandoned Mine Land Inventory System electronic database (Office of Surface Mining [OSM] 2005), except for Slabcamp Run 2, which is from FODC's grant proposal. NRCS figures reflect its full commitment in NRCS (2000), even though an unknown amount has already been spent. Section 319 and OSM figures are from FODC proposals.

2. IMPLAN estimates regional purchase coefficients (RPCs) for each economic sector in each county. RPCs are the amount of local demand that is met by local suppliers and are the basis for estimating economic multipliers. Industries have higher RPCs if most of their demand is met locally and lower RPCs if most of their supply comes from regional, interstate, or international sources.

   IMPLAN multipliers are static snapshots of how an economy functions. Economies, however, are highly dynamic, and IMPLAN multipliers are generally considered to be relevant for 3 to 5 years. Predicting long-term annual benefits would require a dynamic model and more detailed sector data. Dynamic models such as REMI are generally used when a very large project-related expenditure is expected to have a significant immediate impact on the structure of the local economy or key investments are expected to alter economic patterns over time. While a restored stream may change the structure of the local economy in the long run, the change will result from the restored stream and not from the restoration expenditures, making IMPLAN an acceptable model to use for a rough estimate of immediate expenditure impacts.

   The social accounting matrix (SAM) multipliers generated by this analysis account for direct, indirect, and induced spending and employment. The *multiplier* is a ratio between the direct effects of a change in sector demand and the sum of direct, indirect, and induced effects of that spending. *Direct effects* include the first round of expenditures made to carry out the project. *Indirect effects* account for the effects on other local businesses as a result of the initial recipients' need for locally provided goods and services. *Induced effects* describe the demand created in all sectors as a result of any new household income from direct and indirect employment generated by the restoration expenditure.

3. Budget line items for passive and active treatment expenditures were allocated North American Industry Classification (NAIC) numbers based on assumed industry sector matches to determine the relevant IMPLAN model sector RPC (e.g., lime purchases were assumed to match IMPLAN's "mining and quarry" sector). Contacting each good or service vendor to request its actual NAIC number could provide more precise estimates.

4. NRCS contracts must be distributed through open bidding. According to one official working on the Deckers watershed, however, the area's problems with AMD have generated significant local expertise, making it likely that contracts will remain in the area (Yost 2005). This cannot be guaranteed, however. Construction contracts for at least two Deckers watershed projects—Elkins Coal and Coke and Slabcamp 2—were awarded to a Fayette County company.

5. Passive treatment systems are those that do not require significant annual O&M costs and are the preferred type of AMD treatment funded by the AML Trust Fund, NRCS, and Section 319 funds.

6. Each line item in the Slabcamp budget was analyzed to find a sector-specific multiplier and then averaged to reflect a single but mixed economic event. This mix of expenditures was assumed to be standard for all of the passive treatment installation expenditures, so the average multiplier from the Slabcamp project was applied to the total passive treatment expenditure estimate.

7. These calculations assume that construction was performed by a company in the local two-county area. While the Elkins Coal and Coke and Slabcamp 2 projects were actually constructed by a Fayette County firm, it is likely that local firms would construct future projects.

8. The multiplier for passive treatment systems is 1.28, while that for the active systems is 1.17. This is due to the different combinations of products and services that are used to build, install, and operate the different systems. Passive systems will have a greater local impact because more of the goods and services can be provided to the project from the local economy, while active systems draw more on nonlocal resources.

9. It is likely that the land purchase estimated at $250,000 would have additional local economic benefits. However, because there is no information on how or where the recipient would spend those funds, it is impossible to estimate their added local impact. Indirect and induced benefits are therefore assumed to be zero.

10. If funds for O&M expenses were generated locally, then O&M-related benefits listed would have to be compared against benefits forgone from alternative uses of those funds. All other benefit estimates would remain unaffected.

11. Because specific budget information was not available at the time this report was written, the impact estimates were based on multipliers generated by the standardized passive treatment budget analysis.

12. WTP estimates are derived from what is known as the SCM approach to measuring benefits. In this case, this method provides an estimate of what economists call *consumer surplus*. Consumer surplus is the value consumers receive above and beyond the price they pay for a good. Because in the case of Deckers Creek restoration funds would be financed by external funds, respondents receive the good they value at no actual restoration cost. WTP figures, however, can be used as guides for setting access fees or use rates when appropriate or necessary. They can also be used as a guide to determine if investment of public funds will generate an equal or greater public benefit.

The last U.S. Fish and Wildlife economic survey in 2001 estimated the net economic value of an average bass fishing trip in West Virginia at $25 per day. According to FODC (Christ 2005), a restored Deckers would potentially support various types of bass as well as other fish. The average wildlife-viewing trip was estimated to generate $47 of economic value to the visitor. Phaneuf (2002) surveyed anglers to estimate the value of a statewide water quality improvement program in North Carolina and found that anglers' mean WTP for overall water quality improvements was $5.90 per trip.

13. The value of remediation of AMD and garbage but not bacterial problems was estimated to be less than $1 million annually. Respondents expressed a higher value for full restoration than for the sum of each individual restoration benefit. In other words, respondents were willing to pay more for a full stream restoration than a moderate stream restoration. Anglers had the highest value for stream restoration. This underscores the point that benefits from different restoration projects are not simply additive. Restoring aquatic life to Deckers while ignoring bacterial contamination would undermine many users' expected benefits from fishing activities. By the same token, providing a fishable and swimmable stream may encourage more public participation in organized garbage removal activities and discourage dumping and littering along stream banks—an example of future costs reduced or avoided.

14. Economic impact studies of fishing and recreation benefits often attempt to estimate the benefits of new recreation-related expenditures in the community by multiplying the expected number of increased visits by the average spending on a stream-use trip (fishing, boating, etc.). The travel cost method (TCM)—an accounting of the variable costs stream users pay to reach different destinations that are characterized by different attributes—estimates the marginal value of, for example, more fish, better stream quality, increased convenience of amenities, and so on. With these estimates, a demand curve can be derived for users' value of one additional unit of stream quality or one additional unit of fish population.

Using estimates of increased angling use of Deckers Creek to anticipate the impact of increased fishing expenditures or visits in an entire county can be complicated when factoring in substitution effects. Anglers, for example, may make more fishing trips if the opportunity to fish is more conveniently located and surrounded by other activities amenable to a full day of family activity. Alternatively, they may make more trips to a restored Deckers Creek, but these may be trips that the angler substituted in place of, for example, a trip to the nearby Cheat River. Reliably teasing out these behaviors can require extensive and costly surveying even after the restoration, when actual rather than just expected behaviors can be evaluated.

REFERENCES

Cameron, T., and I. McConnaha. 2005. *Evidence of Environmental Migration: Housing Value Alone May Not Capture the Full Effects of Local Environmental Disamenities.* University of Oregon Economics Department Working Papers, 2005-7. Eugene, OR: University of Oregon.

Collins, A., R.S. Rosenberger, and J.J. Fletcher. 2005. The economic value of stream restoration. *Water Resources Research* 41(1): 1–9.

Center for Watershed Protection. 2001. *Economic Benefits of Protecting Virginia's Streams, Lakes and Wetlands.* Virginia Department of Conservation and Recreation, Richmond, VA.

Christ, M. 2004. *Friends of Deckers Creek State of the Creek 2003.* Dellslow, WV: Friends of Deckers Creek.

Christ, M. 2005. *Friends of Deckers Creek State of the Creek 2004.* Dellslow, WV: Friends of Deckers Creek.

Earnhart, D. 2001. Combining revealed and stated preference methods to value environmental amenities at residential locations. *Land Economics* 77(1): 12–29.

Epp, D., and K.S. Al-Ani. 1979. The effect of water quality on rural nonfarm residential property values. *American Journal of Agricultural Economics* 61(3): 529–534.

Hoehn, J., and A. Randall. 2000. The effect of resource quality information on resource injury perceptions and contingent values. *Resource and Energy Economics* 24: 13–31.

Krysel, C., E. M. Boyer, C. Parson, and P. Welle. 2003. *Lakeshore Property Values and Water Quality: Evidence from Property Sales in the Mississippi Headwaters Region.* Report for the Legislative Commission on Minnesota Resources. Mississippi Headwaters Board and Bemidji State University.

Leggett, C., and N. Bockstael. 2000. Evidence of the effects of water quality on residential land prices. *Journal of Environmental Economics and Management* 39: 121–144.

Liston, D., M. Christ, and P. Kasey. 2001. *Remediation of Deckers Creek: A Status Report.* Morgantown, WV: Friends of Deckers Creek.

Loomis, J. 2000. Environmental valuation techniques in water resource decision making. *Journal of Water Resources Planning and Management* 126(6): 339–344.

Monongalia County Parcel Database, Monongalia County Assessors Office. 2005. Accessed July. http://www.assessor.org

Monongalia County Tax Records, Monongalia County Sheriff's Tax Office. 2005. Accessed July. http://www.co.monongalia.wv.us

Natural Resource Conservation Service (NRCS). 2000. *Supplemental Watershed Plan No. 1 and Environmental Assessment for the Upper Deckers Creek Watershed.* Morgantown, WV.

Office of Surface Mining. 2005. *Abandoned Mine Land Inventory System Database.* Accessed July. http://www.osmre.gov/aml/inven/zamlis.htm

Phaneuf, D. 2002. Random utility model for total maximum daily loads: Estimating the benefits of watershed-based ambient water quality improvements. *Water Resources Research* 38(11): 1–11.

Salvesen, D., and H. Renski. 2002. *The Importance of Quality of Life in the Location Decision of New Economy Firms.* U.S. Economic Development Administration. Chapel Hill, NC: UNC Center for Urban and Regional Studies.

Streiner, C., and J. Loomis. 1996. Estimating the benefits of urban stream restoration using the hedonic price method. *Rivers* 5(4): 267–278.

U.S. Fish and Wildlife Service. 2001. *Net Economic Values for Wildlife-Related Recreation in 2001.* Report 2001-3. Arlington, VA.

Yost, P. 2005. Telephone conversation, June 27.

Glossary

Acid Mine Drainage (AMD): AMD is the runoff of metal-rich water flowing primarily from abandoned mines and surface deposits of mine waste, although it can originate naturally from mineral-rich deposits within the earth. AMD occurs when the mineral pyrite (FeS_2), common in coal seams, comes into contact with water and air. The reaction produces ferrous iron, sulfate, and acidity. Ferrous iron oxidizes and produces a precipitate that forms the residue in affected streams known as "yellow boy," while the sulfate and acidity are responsible for lowering the stream's pH. The end result can be streams that no longer support aquatic or plant life if the drainage levels are high.

Averting Behavior (Defensive Expenditures, Replacement Cost): These terms describe the estimated values derived from an analysis of how much it would cost to replace or substitute ecosystem services with new techniques or technology as compared with the cost of protecting the ecosystem.

Benefit-Cost Analysis: In a certain sense, benefit-cost analysis is just a term for one of many decision tools that we use all the time in our daily lives: we consider the option of doing something and think about what it will get us (benefits) and what we have to give up for it (costs). It measures the net gain or loss to society due to a certain policy or project. It provides policy makers with a transparent list of the pros and cons (benefits and costs) of a project to help them decide to support it or not.

Benefit Transfer: Once the benefits and costs are estimated for one project or area, it is sometimes desirable to "transfer" or apply those estimates to another comparable project or site to help estimate costs and benefits associated with it (sometimes this is done to reduce the costs of estimating these figures a second time or because important data for estimation are not available at the comparable site). Benefit transfer has to be done carefully because no two sites are identical, and there are statistical techniques that have to be followed to keep the benefits transfer numbers consistent.

Consumer Surplus: The amount people value a good or service over the price they pay for it. For example, if you are willing to pay 55 cents for a blue lamp at a garage sale and the price is 25 cents, there is 30 cents consumer surplus.

Contingent Valuation Method (CVM): A *stated preference* method that uses surveys to ask people to place a value on a *change* in a certain nonmarket good or service. There is usually some information in the survey about the change, like how a fishing hole is likely to change due to a restoration project (more or fewer fish, better or worse access, etc.). Then the survey respondent is asked how much he would pay for this change. It is similar to *stated choice method*.

Cost-Effectiveness: Meeting a goal with the least costs incurred. Goals could be an ecological or human health standard, for example. It differs from *benefit-cost analysis* in that the benefits are not quantified in monetary terms and compared with or justified by being greater than the costs. When benefits are difficult to calculate or a goal has been set, decision makers use cost-effectiveness to decide *how* to achieve a set goal rather than *whether* to support the goal.

Demand: The quantity of a good or service that a person wants at a given price. The demand curve represents the relationship between various prices of a good and the changing quantities of it that people want under certain circumstances like price changes. The demand curve is based on peoples' preferences and tastes.

Discounting or Interest Rates (Compared with Inflation): Used to calculate the present value of future benefits and costs. Interest rates help to determine how present values should be calculated for different investment decisions. When trying to understand the flow of benefits and costs for environmental decisions, the interest rate is typically referred to as the discount rate. Discounting should not be confused with inflation or the overall increase in prices for a similar bundle of goods and services.

Ecological Economics: New branch of study that is a collaboration among economists and social and natural scientists to address the complex interactions among physical, biological, and economic factors; ecological economic thought does not necessarily rest on assumptions made in welfare economics.

Econometrics: An area of economics that applies data to economic theory to test the strength of the theory or to forecast future behavior. Econometrics relies heavily on regression analysis.

Economic Impact: Analysis for examining market impacts across different business sectors.

Economics (vs. Finance): The study of allocating scarce resources to competing uses. Economics goes beyond simply understanding how money exchanges hands; it examines how different kinds of goods are exchanged and distributed (e.g., money, property rights, positive behaviors, etc.). Economics focuses on scarcity because people always want more than can be produced of almost anything desirable. *See* **ecological**, **environmental**, and **natural resource economics**.

Ecosystem Services: The services that the ecology provides to humans directly or indirectly, such as climate control, water quality management, flood control, etc. There are, of course, services that an ecosystem provides to plants and animals other than humans, but those are biological functions and not usually considered in the definition of ecosystem services.

Efficiency (vs. Equity): Condition when the maximum production is reached with a given amount of inputs and available technology. Neoclassical economics holds that, across sectors of all production in an economy, market prices and free choice should result in the highest production and the highest total social utility possible. At maximum efficiency (*pareto optimality*), changing any allocation of resources results in a net loss of utility. Equity considerations in natural resource allocation are used to explore situations when the goal of maximum production should be reconsidered for purposes of fairness or other social values in the distribution of resources (or in the making of rules for natural resource (NR) markets).

Environmental Economics: Study of how scarce environmental resources are allocated among competing demands. There is a focus on understanding how and why regular markets fail to allocate natural goods to their best social value.

Externality: A benefit or cost imposed on a third party by a market exchange that does not involve them. An example of a negative externality is when a company not only produces and sells a good but also produces pollution that harms people who are not compensated. This pollution is a cost of production, but it is "externalized" if the company is not required or forced to pay that cost. An example of a positive externality is if a wastewater facility must reduce nutrient loadings and it chooses to do so in a manner that also provides green space and wildlife habitat, by implementing riparian stream buffers for example.

Free Rider: An individual who values and who benefits from a public good but strategically chooses not to pay for it. For example, enjoying the restoration benefits of a newly cleaned/restored stream but not donating to the watershed organization efforts to restore it.

Goods: Goods encompass anything that is sought by a person. Goods can be described as *exclusive* or *nonexclusive* and *rival* or *nonrival* to describe how they are able to be enjoyed by consumers. These characteristics of goods have important policy and benefit analysis implications.

Exclusive Goods: A good is exclusive if, because of its physical properties, individuals can be prevented or "excluded" from benefiting from a good or service once produced or easily forced to make a payment on use of that good or service. A small lake can be an exclusive good if access to use it is restricted by membership. A very large lake or, for example, air quality (within a given region) is an *n*-exclusive good because its physical properties do not allow it to be allocated to some and not to others without incurring very large costs. Collecting funds from beneficiaries to pay for maintenance of these goods is difficult without using taxes or moral arguments.

Market Goods: These are both exclusive and rival and allow buyers and sellers to exchange in market efficiently.

Nonmarket Goods: Because these goods cannot be made exclusive (i.e., benefits and costs are not exclusive to buyer and seller), the markets cannot allocate the good efficiently.

Open-Access Goods: Goods are open access if they can become *rival* in consumption (usually due to crowding but also to pollution) and *nonexclusive* (it is costly to exclude selected users from enjoying benefits of the good). A stream is an open-access good because it is costly to exclude users (recreational users/polluters), and high use by some (e.g., polluters) detracts from the ability of others (e.g., recreationalists) to enjoy the benefits produced by the stream.

Rival Good: A good is rival if one person's enjoyment of/benefit from it reduces another person's ability to enjoy/benefit from it. A stream can be rival if one company uses it to dump pollution and then others are excluded from using it for recreation or environmental services. A clean stream or park can be a *nonrival* good because one person or five people could use the good without detracting from the others' ability to use the good (this is limited by the potential for crowding or polluting).

Hedonic Pricing Method: A revealed preference technique that relies on the analysis of price data, often used with home sales, which allows the various aspects of a good to be valued separately. If two houses have the same number of bathrooms, the same number of bedrooms, the same view, the same everything except that one is overwhelmed with a hog farm odor, then we can estimate how that hog smell is valued even though it is never traded in a market.

Marginal Analysis: Examining decisions at the margin or involving small changes.

Marginal Benefit: Additional benefit from one additional unit consumed.

Marginal Cost: Additional cost from one additional unit produced/one additional unit purchased.

Market: Collection of buyers and sellers who exchange goods and services at prices that are determined (directly or indirectly) by supply and demand, by the rules of exchange, and by the additional costs of doing necessary exchange transactions (seeking information about the product, enforcing and monitoring rules and contracts, etc.).

Multiplier: The effect across sectors of the economy that occurs when an expenditure results in a great economic impact because of how the money circulates through the economy.

Natural Resource Economics: Study of allocating nonrenewable and renewable resources that are harnessed as factors of production for market goods and services.

Opportunity Cost: Opportunity cost choosing plan A is the lost value of the option not chosen (plan B). For example, the opportunity cost of using a stream for a waste receptacle could be the value of using the stream for fishing or swimming.

Preference, Revealed: Preferences that are revealed through buying behaviors. When environmental goods or attributes are not traded in markets, we can estimate their values based on prices of related market goods or comparable markets, inferring values for the environmental improvements. For example, the market price of flood insurance could be a revealed preference for the value of protecting an upstream wetland that serves to control floodwaters—both reduce the cost of flood damage.

Preference, Stated: Those preferences we get from surveys by asking people directly how much they prefer or value something. Stated preference can be used to estimate nonuse values.

Present Value: Because people prefer to consume goods and services today rather than in the future, there is a "time value" of money (i.e., sometimes captured in the interest rate). The future value of some investment depends on the interest rate. Present value is the discounted future value. This concept is important when considering decisions that will have benefits and costs that occur in future time periods. If money is preferred today rather than in the future, benefits and costs that occur in the near term will be valued higher than future benefits and costs. Present value is the concept of discounting future values so that all values can be compared on equal terms.

Producer Surplus: The difference between what a firm can sell its product for and the lowest amount the firm would accept for it and still be able to produce it at the same level of quantity and quality.

Public Goods: Goods that are *nonrival* in consumption and *nonexclusive*.

Regression Analysis: A mathematical technique that estimates the statistical relationship between dependent and independent variables. In the travel cost method the dependent variable might be the number of trips a person takes to an area, while the independent variables would be things like his or her income, the distance the area is from home, etc.

Stated Choice Method (SCM): A stated preference model that is survey-based like the *contingent valuation method*. SCM mimics a trade-off by giving the survey respondent a choice between several options that are described by their characteristics. By comparing the options, the data from many such choices can be used to estimate money value of different trade-offs.

Stock vs. Flow: The stock of a natural resource is the asset base from which resources flow. Renewable resources include forests/timber, aquifers/springs, and the like. Nonrenewable resources include coal seams and coal extraction. If the harvest rate exceeds the stock's flow rates, then the value of the stock or asset is decreased.

Substitutability: The degree to which one good can be used in place of another to reach the same goal for the user. A swimming pool is a close substitute for a swimming hole for users interested in swimming; it is a poor substitute for the goal of providing stream fishery and wildlife habitat.

Supply: (1) The quantity of a good or service that a firm or a group will supply at a given price. The supply curve represents the relationship between the market price for a good and the quantities of a good that a firm will supply at those prices. The supply curve is based on the costs the firm faces to make a good or service. (2) The quantity of a nonrenewable resource that exists. (3) The quantity of a renewable resource that is produced (flow) that is the renewal growth rate produced by the stock resource (natural capital) minus the depletion rates.

Total Maximum Daily Load (TMDL): TMDL is the maximum amount of a stressor that a water body can handle and still meet water quality standards. It also allocates the pollutant load among the sources.

Trade-offs: What someone is willing to give up to get something else; this defines economic value.

Travel Cost Model: This is a *revealed preference method* for valuation that is done by estimating how much people spend to travel to a recreation spot; this gives us some idea of what the spot is worth to the person, even though the person might not expressly apply a money value to it if asked. In the travel cost model, we estimate travel expenditures made to visit various sites that differ based on a target amenity (water quality, size of fish, etc.); the differences in peoples' willingness to invest in travel costs to seek out better amenities helps us value those amenities.

Utility: The well-being benefits derived from a good or service.

Value: A good's economic value is a measure of how much it matters to people and how it affects an individual's perceived or actual well-being. Well-being says something about our satisfaction level from an action or activity. There are important differences in components of value.

> **Bequest Value:** A type of nonuse value, bequest value is the value we place on a thing because we want it to be passed down to future generations.
>
> **Economic Value:** A good's economic value encompasses its financial value plus all of the other values to society that are not traded in a market. The expensive house by the river (mentioned in the discussion of the hedonic pricing method) may have a negative *economic value* if building it destroyed valuable wetlands

that used to help control floods and provide habitat for wildlife that others used to enjoy in the neighborhood. The value of the house then equals its financial value minus the cost (plus the negative value) of flood damage minus the cost to others in the neighborhood who enjoyed the wetland and the creatures that rely on it.

Existence Value: A type of nonuse value, existence value is a value we place on an environmental amenity simply because we want it to "exist." Many people value what are called *charismatic mega fauna* (lions, tigers, whales, elephants, etc.) not because they use them or even ever plan to see one in the wild but because they value their existence.

Financial Value: A good has financial value if it is or can be exchanged in a market for money. The opportunity to live by a river has a positive financial value. We know this because in the real estate market, a riverside house costs more than an identical house sitting three blocks back from the river.

Nonuse Value: The value encompassed in a good that we enjoy but do not necessarily use. These values are real, but we have a hard time quantifying them; they are things like the value we hold in knowing a thing exists, valuing someone else's use of a thing, or knowing we might someday want to use or enjoy the thing, if not now. Nonuse values can be potentially very high.

Option Value: The value to know something will exist for use in the future. Some people hold an option value for coal in the ground (and other natural resource reserves) knowing that using it now will not allow it to be used later.

Use Value: The value we place on things because we actively use and enjoy them. Indirect use value is the value we place on something that, while we might not use it, we use something that depends on that thing. We may have indirect use value for mayflies if we fish, for example.

Water Quality Standards (WQS): WQS are the foundation of the water quality-based control program mandated by the Clean Water Act. WQS define the goals for a water body by designating its uses, setting criteria to protect those uses, and establishing provisions to protect water quality from pollutants.

Welfare Economics: Study of how individuals or firms try to maximize their well-being (utility or profit) through their choices, with the assumption that, collectively, those choices result in an optimal production and allocation of goods in the market and society.

Willingness to Pay (WTP): Refers to the value of a good defined by what a person would be willing to pay for an item that is not offered in the marketplace (their buying price). The good may be protection of a fish in a stretch of stream or restoration of a destroyed fish habitat. This value can be estimated by surveys and other research tools such as *revealed* or *stated preference methods*.

Willingness to Accept (WTA): Refers to the value of a good defined by what a person would be willing to accept in exchange for its loss—their selling price (even though it may not be theirs to sell). The good may be headwaters that are buried by mountain top removal or loss of a community park. This value can be estimated by surveys and other research tools such as *revealed* or *stated preference methods*.

Index